S0-BWX-387

THREADS OF LIFE

THREADS
OF LIFE

Genetics From Aristotle to DNA

AARON E. KLEIN

Drawings by Robert Michaels

PUBLISHED FOR THE AMERICAN MUSEUM OF NATURAL HISTORY
The Natural History Press, Garden City, New York

CREDITS

The Natural History Press, publisher for The American Museum of Natural History, is a division of Doubleday & Company, Inc. The Press is directed by an editorial board made up of members of the staff of both the Museum and Doubleday. The Natural History Press has its editorial offices at The American Museum of Natural History, Central Park West at 79th Street, New York, New York 10024, and its business offices at 501 Franklin Avenue, Garden City, New York 11530

For Eric and Jason

CONTENTS

PROLOGUE

This book is an account of some of the men whose work led to one of the most profoundly important series of discoveries in the history of human achievement. These discoveries will have more effect on the future of mankind than all of the wars ever fought.

Most of the men involved never thought their work particularly momentous in an historical sense. As scientists they pursued their own search for knowledge, never knowing that it would blend with other lines of research to point out an essential unity of the work of all scientists. The unity had always been there, but it had been lost in the necessary details of the work of scientists in particular fields or "disciplines" of science.

A climax in this work has been reached in the past decade with a virtual explosion of scientific work following the clarification of the structure of a molecule. Much of what had eluded scientists, especially biologists, has been found in the chemistry and physics of this now well-known molecule—DNA.

Where does this story begin? Actually, it begins with the first glimmer of wonder in the first creature that could be called a man. For our story we will go back to the middle of the sixteenth century when man was emerging from a long intellectual nap. The men who led the awakening were not always popular. Those who would wake us from the comfort of sleep are not always gratefully received.

1. AT THE PORTALS OF THE MAZE

The first decades of the sixteenth century were perilous times in Europe. The Christian church, long a unifying force, had split, foreshadowing long periods of war and upheaval. In the face of the end of Christian unity, the Turks were slowly advancing from the east. The Turks, led by Suleiman the Magnificent, may well have conquered all of Europe, had they not been narrowly defeated in an attempt to take Vienna in 1529.

In such times a man needed comfort in familiar, solid things. Facts that are known to be unalterable and forever true always have been a good source of comfort. It was comforting to know that one lived in the center of the universe, and that the sun, moon, and all the stars revolved about him. As a student, it was comforting to know that all you had to do to get a good grade was to memorize enough Aristotle to satisfy the teachers. Every man found some comfort in the knowledge that all the living things in the world had been made in one great act of creation and that there were only so many kinds of living things that would always remain as they were created. Man could take further comfort in the fact that he had been made in a special act of creation and was set apart and above the other living things in the world. Then, as now, those who would destroy a man's comfort with new thoughts and ideas were not likely to win popularity contests.

If a man wanted to become learned, he could read Aristotle. It was widely believed that almost everything that was worth knowing could be found in the works of this philosopher. This was especially the case as far as natural science, the study of living things, was concerned.

Aristotle has been recorded in history as the best known of the *natural philosophers,* who flourished in Athens around 300 B.C. These men believed that there was a natural order to the universe and its ultimate control was in the hands of the various gods. The natural philosophers reasoned that although the order of the universe might be controlled by the gods, knowledge of this order need not be limited to the gods. Man could, if he gathered enough information through observation, completely understand the natural order of the universe.

Aristotle did not limit his activities to natural history. His interests included mathematics, astronomy, and politics. He did not experiment in the manner of present-day scientists. His method was to observe as much as possible and draw conclusions from what appeared to be logical and reasonable. He wrote generalized accounts of the dissection of many animals and made a detailed study of the development of the chick.

Many of his observations are still accepted today as essentially correct. As might be expected, much of what he concluded is just not right in view of today's knowledge. For example, he concluded that the brain was a cooling organ for the body. He believed that veins carried air. He probably arrived at this last conclusion because the blood had drained from the veins of the dead animals he dissected.

Aristotle did a good deal of teaching. One of his students, Alexander the Great, conquered much of the known world and in so doing caused the spread of Aristotle's writings to Asia and

North Africa. Alexander was somewhat egotistical and had a habit of decreeing that cities be named after him, and, in at least one instance, after his horse. The best known Alexandria is the one in Egypt. A great center of scholarship and learning flourished there for hundreds of years.

The Romans knew of Aristotle, but most of the rest of Europe never heard of him until long after the Roman Empire ceased to exist. After the decline of Rome, Europe fell into a state of general disorder that has come to be known as the Dark Ages. During this time, ignorance and superstition were widespread. Education and the gaining of new knowledge were practically nonexistent.

But at the same time, centers of learning flourished in areas of the Middle East that had been conquered by Alexander. Alexander died a young man, but the works of the natural philosophers had found fertile ground in which to grow. The Arabs developed a system of writing numbers, and invented algebra. In Alexandria, a great library and museum were built. Many of the works of Aristotle were translated into Middle Eastern languages.

The astronomer, Claudius Ptolemy, lived in Alexandria from A.D. 85–163. Ptolemy, drawing on his own observations, plus the works of Aristotle and Babylonian astronomers, established an idea of the movements of the stars and planets. The essential feature of Ptolemy's astronomy was that the earth was the center of the universe and that the sun, moon, planets, and all the stars revolved about the earth. This idea was to be accepted as an absolute, unalterable truth for another 1600 years.

Meanwhile, back in Europe, the natives began to stir. Christianity had spread rapidly through Western Europe after the decline of Rome. The Church provided leadership and some degree of unity. By 1095 economic conditions and communications had improved to the point where all of Western Europe could unite toward a common goal. The goal was the Holy Land, where Christianity had originated. Pope Urban II issued an appeal that all local wars should stop and that all Christians unite to take

Jerusalem back from the infidel Moslems. The response to the
Pope's appeal was enthusiastic. Thus began the First Crusade.

There were many more Crusades over the next century and a
half. Jerusalem was taken and retaken by Christians and Moslems.
One result of the Crusades was the devastation of the countries of
the Middle East. Another was the finding of the works of Aristotle
and other scholars by Western Europeans. The Arab Moslem
lands fell into a decline from which they never have completely recovered. Western Europe gained the seed that was to grow into
what may have been the greatest flowering of knowledge the
world had ever known.

The works of Aristotle were translated from Greek and Arabic
into Latin and the translations widely distributed. The introduction
of Aristotle's works into Western Europe was like a blood trans-
fusion that revived a dying man. Some of his work was known in
Europe before the Crusades, especially in Spain. After the
Crusades, however, Aristotle's fame spread rapidly throughout all
of Western Europe.

Man is by nature a curious animal and the desire for knowledge
remained in the minds of many men even during the long centuries
of the Dark Ages. Literate men eagerly read the works of Aristotle,
Ptolemy, and other Greek and Middle Eastern scholars. It is somewhat ironical that the works of Aristotle led to another decline
in the search for new knowledge. Gradually, Aristotle came to be
accepted as the final word on many matters. From time to time,
to doubt Aristotle was heresy, treason, or both. The universe as
described by Ptolemy came to be an article of faith, and to speak
against it could result in excommunication from the church.

A Franciscan monk of the thirteenth century named Roger
Bacon had the courage to question the accepted order. He proposed
that new knowledge be gained by experiment. He respected
Aristotle but maintained correctly that much of it had been badly
translated. In one of his books he wrote, "If I had my way, I
should burn all the books of Aristotle, for the study of them can
only lead to a loss of time, produce error, and increase ignorance."
If Aristotle had been alive he might have agreed with Bacon. For
his efforts, Bacon was confined to his monastic cell for long periods

of time, had his every movement watched when he was not confined, and was very nearly excommunicated. After his death, his
writings were tucked away where no one could find them. He was
not generally known to historians until some 500 years after his
death.

It was not the memory of Roger Bacon that troubled the comfort seeker of the sixteenth century. The comfort destroyers were
very much alive and would soon make their presence known. By
the sixteenth century the printing press was in wide use. More
people could read in the sixteenth century than had been the case
in Roger Bacon's time. There were many people who saw the
contradicton between what they read in Aristotle and what they
observed with their own senses. A new world had been discovered
on the other side of the Atlantic Ocean. Reports of strange animals
and plants in the new world were causing some people to have
doubts about the unchangeability of living things. Many people
realized that there was much to be learned, much more than
could be learned from the dusty volumes of Aristotle.

It is, of course, impossible to specify a date for the reawakening
of scientific inquiry in Europe. The year 1543, however, is worthy
of mention. In that year, two men, one an anatomist and the
other an astronomer, published books in which they expressed
their doubts about the established order. Neither of these men
was a fire-breathing revolutionary who set out deliberately to
change the world. But intentionally or not, that is exactly what
they did.

Andreas Vesalius was a professor of anatomy at the University
of Padua in Italy. Many universities had been started in Europe,
especially in Northern Italy, as part of the general reawakening
of European arts and sciences, called the Renaissance. Most of
these universities had departments of medicine. These medical
schools had their own "Aristotle" in the writings of Galen. Galen
was a Greek physician who lived about A.D. 140–200. His works
had been carried to Europe in much the same way as the works
of Aristotle. As with Aristotle, in many matters, the writings of
Galen came to be accepted as the final word in human anatomy.

In a typical scene in one of these medical schools one would see the professor sitting on a raised chair reading Galen. Below him sat the students listening and watching the dissection of a cadaver (dead body). The professor would not dissect. Such unpleasant work was considered to be beneath his dignity. The actual dissection was carried out by an operator, usually a barber, under the direction of the professor's assistant.

In the course of the dissection if anything was found that did not agree with Galen, it was dismissed as a "freak of nature." This was so even if the freak of nature had been observed in every cadaver dissected. Many students wondered why the freaks persisted, but to ask too many questions was dangerous. Many professors noted the contradiction between reality and Galen, but most were afraid to speak out for fear of losing their positions.

Vesalius revered Galen to the point of adoration. Galen had been remarkable for his time, and was one of the first people to point out the importance of studying the human body. Galen, however, had used an animal called the Barbary ape as a dissection specimen. In applying his observations of the Barbary ape to humans, he was in error. Vesalius felt that these errors should not be taught as fact.

Vesalius shocked the other anatomists by performing his own dissections—the barbers were given the job of sharpening the knives. He infuriated the other professors by basing his instruction on what he saw in the dissections and not entirely on Galen.

His book, *De Corporis Humani Fabricia* (The Fabric of the Human Body), had many beautifully drawn anatomical figures in it. Upon publication of the book, many traditional anatomists were even more furious with Vesalius. Many said that he should be removed from his position and be forbidden ever to teach again. Within a year, however, Vesalius had a large following of students. He was so completely accepted that he received more offers of professorships than he could possibly accept. Anatomy became a rapidly advancing science.

The astronomer Nicholas Copernicus had been concerned that the mathematics of the Ptolemaic system was too complicated. He felt it should be simplified. In the course of his work he found that the movements of the heavenly bodies could be more easily explained mathematically if the earth and planets revolved about the sun. He went on to say that if the earth moved, then the apparent motion of the sun and stars was an illusion caused by the motion of the earth.

Not wanting to upset anyone, he had a friend write a lengthy preface to his book. In the preface, it was explained that he really was not trying to change anything, but was only offering a simpler explanation for what already existed. The impact of his work was slow to develop. He died the same year his book was published. Others who followed him, notably, Galileo, Kepler, and Newton expanded on Copernicus and said what he had been afraid to say. Galileo, however, was forced publicly to deny what he believed to be the truth.

Copernicus could not be suppressed for very long. By the eighteenth century the scientific evidence was conclusive. The earth was not the center of the universe. A creed that had been accepted for over 2000 years had been shattered. If the methods of science could bring down one "unalterable truth," others could and would be brought down.

In contrast to the quick acceptance of Vesalius, the astronomy of Copernicus was to be debated furiously for over a hundred years before it gained respectability. Why was this the case? Most anatomists were ready to accept Vesalius. He had the courage to point out what most anatomists knew, but were afraid to admit. It was relatively easy for Vesalius to prove a point. Galen had written that there was a bone in the heart. All one had to do to see that there was no bone in the heart, was to get his nose out of the Galen manuscript and dissect a real heart.

Copernican astronomy was not so obvious. In order to understand Copernicus, one had to be fairly knowledgeable in sophisticated mathematics. Even if the mathematics could be understood, living in the center of the universe was just too sweet a thought to give up without a fight. In addition, Copernican astronomy was contrary to accepted religious ideas, whereas very few people considered the shape and position of organs to be particularly sacred.

Anatomists of the time did experience difficulty in obtaining dissection specimens due to church restrictions.

The new age in human thought that was dawning has been called the Age of Enlightenment. The advances of this age were not limited to science, but also included art, music, literature, and politics. The work of two men, Francis Bacon (1561–1639) and René Descartes (1596–1650), stimulated the scientific revolution of the Enlightenment. The full impact of the work of these men was delayed by the disastrous Thirty Years' War that devastated Europe from 1618 to 1648. The Enlightenment was primarily a movement of the eighteenth century. At first, it had its main centers in England and France. In Germany, which had been practically flattened by the war, the Enlightenment was even further delayed.

As his namesake had done some 300 years earlier, Sir Francis Bacon wrote that man should go to nature for new knowledge rather than to the dusty manuscripts of Aristotle. Bacon's main contribution to the scientific revolution was a book called *The New Atlantis*. In this book he proposed a fantastic utopian world. His dream world would be filled with "Temples of Science" where men could spend all their time pursuing new knowledge. There were no actual temples, but men did get together to organize various "scientific societies." These societies enabled a free exchange of ideas. There were a few such societies before Bacon, but after *The New Atlantis* they sprang up everywhere.

By the second half of the eighteenth century, science was definitely the "in" thing among the important educated people of Europe. The Royal Society for the Improvement of Natural Knowledge, chartered in 1662, had among its members Benjamin Franklin, King Charles II, and Sir Isaac Newton. Participation in scientific work was almost entirely limited to wealthy and titled gentlemen in the seventeenth and eighteenth centuries.

The bewigged gentlemen in knee britches, stockings, and ruffled shirts of the time looked forward to the meetings of the societies. At the meetings they would watch demonstrations of static electricity, mechanical motion, and examine the latest new specimens of animals and plants from America and Africa. They eagerly looked at the craters of the moon through telescopes and at tiny, strange creatures through microscopes. They would argue passionately over matters of which they knew little or nothing. Yet, in the process, they learned more and added to man's fund of knowledge. They were no longer bound to accept without question the words of a man who had lived 2000 years before them. Never before had there been such intense intellectual stimulation and inquiry.

The Royal Society began to publish the minutes of the meetings in a regularly printed pamphlet called *Philosophical Transactions*. This was the first scientific journal. Other scientific societies soon adopted the idea. These journals were sent all over Europe and to the Colonies in America. Interested people could now participate in scientific work without actually attending the meetings. The scientific revolution was becoming world-wide. A chain of discovery had begun that eventually would change the world of the splendidly attired men of the societies.

Where Bacon was concerned with the proper place and atmosphere for scientific inquiry, René Descartes proposed the method. Descartes wrote a book entitled *Discourse on Method* in which he reported on the procedures he used to carry out investigations. Among other things, Descartes proposed that a scientist should begin with the simpler aspects of his problem and proceed to the more complex parts. He cautioned that a scientist must maintain an open mind in his observations and not be influenced by personal prejudices in drawing conclusions. Descartes reasoned that since the universe was infinite, the knowledge that man could obtain was infinite.

Under the influence of Descartes, thinking men came to the realization that there was no limit to the accumulation of knowledge. Scientific inquiry could not only extend outward to new fields, but could extend inward to study in more detail what was generally known. Scientists began to specialize. The Greek natural philosopher tried to study the entire natural universe. The new scientist began to concentrate his efforts in a particular field such as astronomy, physics, zoology, or botany. This trend to specialization intensified and continues today. The chain of discovery would go in many different directions, somewhat as in a maze. And, as in a maze, there would be many wrong turns and dead ends. In time, the paths would emerge to reveal a unity of life that even the natural philosophers of Athens could never have imagined.

The natural scientists, both of the zoologic and botanic variety, had enough to keep them occupied in the eighteenth century. The explorers were bringing back all kinds of animals and plants from lands across the seas. The naturalists eagerly turned their attention to these specimens and were immediately troubled by an old question. Just what is a "kind" of animal? Even primitive men knew that there were different kinds of living things. It is obvious that a horse is one kind of animal and a dog another, but are a Dalmatian and a cocker spaniel the same kind of animal or different kinds of animals? Is a Jersey cow the same kind of animal as the Auroch from which it descended?

The concept of species for a particular kind of animal was developed in Aristotle's time, but the naturalists of the Enlightenment could not explain what they meant by the word species any better than could Aristotle. The question still troubles biologists today. A species is usually thought of as a population of living things that interbreed with each other, producing offspring of the same species. This definition of species is far from adequate, for under certain conditions, organisms that are considered to be different species do interbreed.

In the seventeenth and eighteenth centuries most people still believed that the number of species was fixed. That is, that all species had been created at one time and no new ones had ever come into existence. Many prominent naturalists believed in the fixity of species, if for no other reason than that establishing species was difficult enough without the complication of species which changed all the time.

It was also obvious that, for example, a lion and a tiger seemed to be related in some way. Then, too, man was not like a tiger, but somehow he seemed to be more like a tiger than a frog. The gentlemen scientists argued that there had to be some kind of "natural" way of grouping living things. There also had to be a better way of naming living things. The sea captains were bringing thousands of specimens to the naturalists for their consideration.

There was a great need for someone who could establish some kind of order out of the chaotic array of specimens. The gentlemen of the scientific societies did not have to wait for him very long. He was one of the first of the few giants of thought who put together the thoughts of many.

Carl Linnaeus was somewhat of an exception among the scientists of the early Enlightenment. He was neither French, English, nor wealthy. He was Swedish, the son of a minister. In 1717 when Linnaeus was born, the influence of English science was rather strong in Sweden. Those who could afford it, through possession of money or title, sent their sons to study in English universities. Carl Linnaeus attended Swedish universities and was grateful to be able to attend a university at all.

He started as a student of medicine, but much to the anger of his father, he showed little interest in medical studies. His father feared that he would never amount to anything. The elder Linnaeus would have been very much annoyed had he known his son was being distracted from medicine by some French botany books. This distraction was indeed fated to keep young Linnaeus from ever amounting to anything as a physician. It was also to make Carl Linnaeus a legend in his own lifetime.

Linnaeus read the French botany books, and wrote a little paper on their contents. In this paper he stated his belief that the sexual parts of the flowers of plants could be the basis of a system of classification. The professor of botany at the university was so impressed with this piece of work that he made Linnaeus his assistant.

In 1732 Linnaeus was appointed plant collector of Uppsala Academy. He soon set out across Lapland with his collecting cases, boots, microscope, and books. He walked all the way to the Arctic Ocean. He was almost devoured by mosquitoes, almost drowned several times, and suffered alternately from heat and cold. In the course of these hardships, he collected hundreds of specimens of animals and plants, many of them previously unknown.

After his journey through Lapland, Linnaeus traveled to England, Holland, and France and presented his specimens and ideas at meetings of various scientific societies where members, especially in England, were very much impressed with the young Swede. Linnaeus was by no means the first person to propose a system of classification. Many who came before him, notably the Englishman John Ray who died two years before Linnaeus was born, had worked on the problem. The time was very opportune for Linnaeus. There was a mania for naming and grouping the thousands of new plants and animals that were swamping the naturalists. The scientific world soon turned to Linnaeus as the supreme authority in matters of naming and classifying.

Inspired by Linnaeus, young naturalists traveled to unexplored parts of Africa and America to gather specimens they hoped to present to the master. Many of these naturalists were killed by pirates or disappeared in the wildernesses as probable victims of native aborigines. The binomial (two-name) system of naming developed by Linnaeus provided added incentive for many of the collectors. It was possible that a collector could have the honor of having a new life form named after himself.

This binomial system of naming living things is still in use today. Linnaeus maintained that at least two names were necessary to designate a species. For example, the common dog would be called *Canis familiaris,* the common cat would be designated *Felis domesticus.* The first word of the name indicates the genus and the second name places the organism in a species. The words are usually Latin, and in most instances, describe something about the organism. *Canis* is from the Latin word for dog, and the second part of the name indicates that it is a very familiar sort of creature. In some cases one or both parts of the name may be a Latinized version of the name of the person credited with discovering the organism. *Oenothora lamarckiana,* the evening primrose, a case in point, is named after the scientist Lamarck.

Linnaeus went beyond just naming and attempted to develop a system of classifying organisms into groups that would express the relationships of the organisms. Similar species were grouped into genera (singular, genus), genera into orders, and orders into classes. The class was the largest category, and contained the largest number of organisms. As one went down from class to species, the number of organisms in each group was smaller, and they had more in common with each other. The system used today is much the same. A few more groups (phylum, family, etc.) have been added. The confusion over where to place what and why is still with us, as it was in the time of Linnaeus.

Linnaeus was influenced by a good many things. He firmly believed in the fixity of species. But there was something else a good deal more subtle than the specific work of those who came before him. Since the time of the Greek natural philosophers, there had arisen a concept of the general order of living things. The idea was that when living things were created, they were made in an increasing order of complexity. This "scale of being" or "ladder of nature" had man at the top of the ladder with the other creatures below him. This scale was pictured to be a straight ladderlike thing with "improved models" on each higher rung of the ladder. Linnaeus, as almost everyone else in his time, believed that every creature had a fixed, unchanging place on the ladder. He not only held to the fixity of species, but also believed that no new species had come into existence since the Creation. He also held that no species had become extinct since Noah's flood. The scale of nature concept supposedly expressed the unchanging place and existence of every living thing. Ironically, the concept stimulated a chain of thought that was to contribute to the overthrow of the idea of fixed species.

Linnaeus wanted to name and classify every living thing in the world. He was completely absorbed in the sheer wonder of creation. To him, the world was a great garden in which living things were lying in wait to be named by Linnaeus. His best-known work was the *Systema Naturae*. In this book he explained his system and catalogued the species he named. In his lifetime he wrote ten editions of *Systema Naturae*, each edition larger than the previous one.

When Linnaeus died, his funeral was more like that of a king than a botanist. He was more of a legend than he was a real person. Along with the adoration and honors that came to him in his lifetime there came doubts and criticism. His system based on the sexual parts of flowers was criticized as "artificial." He readily admitted this, but maintained that such systems were necessary until a "natural" system was found. We are still looking.

Far more serious to Linnaeus was the growing doubt of the fixity of species. Strong evidence against the fixity of species existed in farms and gardens. Men had been selectively breeding animals and plants since the beginnings of civilization. The wheat that was grown on farms in the eighteenth century bore no more resemblance to the wheat of biblical times than a tabby cat to a saber-toothed tiger. If new kinds of animals and plants could be produced on the farm, why couldn't new kinds be produced in nature?

At times, a new variety of plant would appear spontaneously in a farmer's field. Such new varieties were called "sports," and no one could explain what made them appear. The pattern of distribution of living things was difficult to explain, if all species were supposed to have left the Ark at the same geographical location. Why were some species found only in certain parts of the world? Fossil remains of animals and plants had been found. Living examples of these fossils could not be found anywhere. Was this not evidence that some species had become extinct? The belief in the fixity of species was as strongly entrenched as the universe of Ptolemy had been. Many men such as Linnaeus would not admit that the growing mass of evidence to the contrary was significant. Others were impressed by the evidence, but were afraid to express their beliefs. As in all ages, however, there were those who had the courage to say and write what they believed to be the truth.

George Louis Leclerc, the Comte de Buffon, shared with Linnaeus the same year of birth. The circumstances of the two men, however, were quite different. The Comte de Buffon came from a wealthy and distinguished French family. As a man of leisure and wealth he had the time, resources, and inclination to devote his life to science. Unlike many of the gentlemen scientists of his time, he also had a great deal of ability.

As a scientific investigator he did not work very hard. This was in contrast to Linnaeus, who was a painstaking and energetic worker. The ideas of Buffon, many of which were in direct conflict with Linnaeus, excited the imagination of many men, especially in France. Much of his popularity probably was due to his brilliant style of writing. He was more popular in France than in England possibly because of the loss of sparkle his writing suffered in translation. That anyone read Buffon's work at all is a tribute to his writing skill. His great work, *Histoire Naturelle* (Natural History), consisted of no less than forty-four volumes written over a period of fifty-five years. The last volume was completed by some of his followers after his death.

In his massive work, Buffon tried to include all known scientific knowledge. Of course, he could not possibly succeed in this effort, but in the process he did something far more significant. Scattered through his forty-four volumes are references to ideas that were as revolutionary to natural science as the ideas of Copernicus were to astronomy. He was one of the first of the well-known naturalists to suggest that species were not fixed in number and form. Furthermore, he was brave enough to acknowledge his statements. Other naturalists had suggested that species were not fixed, but had done so anonymously or under false names. The atmosphere may have been freer during the Enlightenment than was the case during the Dark Ages, but to suggest that species were not fixed was still somewhat dangerous.

Even Buffon did not shout his ideas from the rooftops of Paris. The reader had to be rather clever to understand what Buffon was trying to say. He denied in one volume what he said in another. Much of this self-contradiction may have been due to lack of any real knowledge rather than to fear of the established system.

Buffon had observed the production of new varieties in the farms and gardens. He had observed fossils and recognized that they represented the remains of species which no longer existed. Buffon never attempted to bring down the idea of the all-inclusive creation. He did say that some of the original forms of the Creation no longer existed. In one volume he wrote that "families of living things are conceived by nature and produced by time." The meaning of this rather lyrical statement is perhaps clearer today than it was in Buffon's time. It is rather easy to say, looking back from our own time, that Buffon said the originally created organisms somehow had changed over long periods of time to produce the organisms of today. To have understood the statement in Buffon's day would have required a much greater concept of time than was generally available in the minds of men.

It was widely believed that the earth had been created in 4004 B.C. The somewhat less than 6000 years thus afforded the earth was hardly enough time for much gradual change of organisms to have taken place. Other gentlemen of the Enlightenment, whose interests were in rocks and the sediments of river beds were to provide the naturalists with the earthly time they needed to propose the evolution of living things with certainty and confidence.

Men in so-called primitive cultures had more of an idea of the great age of the earth than did "civilized" men in Europe in the seventeenth century. The English naturalist John Ray expressed the generally held ideas when he wrote of some observations he made in 1663. A forest that had been buried on the bottom near the shore had become exposed to dry land. Ray noted the hundreds of feet of sediment that rivers had deposited in this area. He commented that it was difficult to understand how so much sediment could have been deposited in the scant 5500 years that the earth had existed. He also noted that if mountains had not been here when the world was created, then the world was probably a good deal older than it was generally thought to be.

As some men had turned to the heavens and the world of living things in their search for knowledge, others had turned to the earth itself. One of the first observations of the structure of the surface of the earth, was that certain rocks occur in layers or *strata*. Quite frequently, the fossil remains of animals and plants were found in the strata. When the remains of salt-water organisms such as shelled molluscs were found hundreds of miles from the nearest ocean, explanations were in order. The traditionalist explained away such finds by reminding everyone that there had been a great flood. Others maintained that finding marine fossils on dry land was an indication that the earth had gone through tremendous changes. Those who held the latter view had different opinions as to how these changes occurred and how long they took to occur.

Buffon in his *Theorie de la Terre* (Theory of the Earth), published in 1749, commented that the strata "were not formed in an instant, but were gradually formed over long periods of time." Buffon went so far as to suggest that the "long period of time" was about 70,000 years. Even this very conservative estimate of the age of the earth met with opposition from those who felt that the planet had been made for man, and that man had always been on it. Other men offered their ideas on the origin of the strata. A group called the "Neptunists" proposed that the strata were sediments that had settled out of a vast sea that once covered the earth. Another group called the "Plutonists" said that volcanoes had deposited the sediments that made up the strata. These two groups spent so much time fighting and insulting each other that they contributed very little to solving the problem of the history of the earth.

The bickering of the Neptunists and Plutonists, the fanciful speculations and the mysticism that characterized geological investigation in the seventeenth century delayed the emergence of geology as a real science. The contributions of three men, James Hutton, William Smith, and Baron Cuvier stand out during this period. Even they failed to complete the picture.

James Hutton, the oldest of the three, came closest to seeing the picture. Hutton saw the earth as being constantly built up and worn down in an endless cycle of gradual accumulation of sediment, volcanic activity, and erosion. He called his picture of the earth a "beautiful machine." His view of constant change and of the forces that bring about these changes is very much like modern geological theory. Hutton published his findings in 1795. His work, well received at first, was eclipsed by the views of those who would compromise with the traditional view.

William Smith, an engineer and surveyor, observed that particular strata always had the same kind of fossils. He was able to identify strata in different locations by the fossils they contained. Neither Hutton nor Smith were naturalists. The third man was a naturalist. His view of the history of the earth was to confuse a whole generation of naturalists.

Baron Cuvier spent most of his time studying the bones of long dead animals. As such, he has been called the "father of paleontology," the study of fossils. As far as fossil bones were concerned, he was a genius. He maintained that from a single bone or feather he could reconstruct an extinct animal by comparing the relic with existing animals. This sort of thing is done in museums today and is no less remarkable than it was in Cuvier's day.

In the course of his fossil studies Cuvier became the champion of that view of the earth's history known as catastrophism. According to the catastrophists, the earth had undergone a series of gigantic upheavals from the Creation to the flood of Noah. As a result of each cataclysm, all living things were wiped out and new ones were created after each catastrophe. The strata were seen as representative of periods between cataclysms.

Cuvier's theory was an excellent means of keeping the traditional idea of the Creation while still accounting for the strata and the bones of creatures that no longer existed. Hutton's "beautiful machine" was the basis of the view called uniformitarianism. As long as Cuvier enjoyed prestige, the uniformists were "out."

Naturalists such as Erasmus Darwin and Lamarck were to grope for answers about changing species while under the influence of the catastrophic doctrine. Not until some sixty years after Hutton's death was a man named Charles Lyell to vindicate the "beautiful machine" of uniformitarianism and establish geology as a true science. The naturalists of the nineteenth century would have all the time they needed to consider the problems of the evolution of species.

By the end of the eighteenth century there was a group of naturalists who were convinced that species were not fixed in number and form. Just how new species came into being was the primary concern of these naturalists. Erasmus Darwin was far more blunt and specific in his writings than Buffon. He stated that change is a characteristic of nature. He cited the example of a caterpillar changing into a butterfly, and teased the traditionalists by asking them if butterflies had been created as caterpillars or butterflies. He pointed out the changes that were brought about by selective breeding and examples of such animals as the snow hare whose fur changes to white in the winter and is thereby concealed in the snow. He commented on the similarity of structure among many animals and suggested that there was an overall relationship.

Erasmus Darwin proposed that changes are brought about by changes occurring in the environment or surroundings of the organism. In stating that changes were induced by the environment, he implied that these changes were passed on to succeeding generations. This idea has come to be known as the inheritance of acquired characteristics. It is more closely associated with another scientist—Lamarck. The ideas of Erasmus Darwin were to impress a yet-to-be-born grandson. This unborn grandson was to completely overshadow Erasmus Darwin in the histories of science that were waiting to be written.

Jean Baptiste de Monet Lamarck is the name usually associated with the idea of the inheritance of acquired characteristics. Lamarck had many other misfortunes. He was married four times and had to support a large collection of children on a rather small income. His outspoken ideas on the changeability of species generated the dislike of Cuvier. In the scientific world of the late eighteenth century it was not exactly helpful to one's career to be on the wrong side of the great Baron. A botanist most of his life, Lamarck was forced at the age of fifty to change to zoology to keep his job at a natural science study center called the *Jardin de Roi* (Garden of the King).

Despite his misfortunes, he accomplished a great deal during his career as a naturalist. When he was seventy-one, he started a massive work on invertebrate zoology. He completed it in eight years and became established as an authority on invertebrates. He was probably the first zoologist to show that spiders are not insects. Since he was a well-known personality, keeping his head during the turbulent years of the French Revolution was no mean accomplishment.

Lamarck invented the word "biology." That is, he invented it, insofar as the present usage of the word is concerned. The term biology grew out of his idea that there was a "natural sequence" of organisms. This was essentially the scale of nature idea that influenced Linnaeus and others. However, while the scale was fixed to Linnaeus, it was more like an "escalator" to Lamarck. Lamarck theorized that lower forms of life "strive" to improve and give rise to forms that occupy a higher place on the ladder. To Lamarck, the boundaries between species were indistinct. He advocated that all living things should be studied together rather than as separate species of animals and plants. It was for this unified study that he proposed the word biology.

To Lamarck, the constant emergence of new species "was the accomplishment of an immanent [sic] purpose to perfect the creation." Lamarck was, and still is, ridiculed for saying that organisms strive to improve. Erasmus Darwin used the word strive in his writings, but he has somehow escaped the abuse that has been Lamarck's legacy. Both Lamarck and Erasmus Darwin were convinced of the mutability (changeability) and emergence of new species. It was the "how" they were groping for when they spoke of striving. Fertilization of egg cells by sperm cells was not understood in Lamarck's time. Nothing at all was known of the mechanics of heredity. Lamarck's attempts to explain the how of changing species was somewhat like looking for an object in a completely dark room. Lamarck was ahead of his time, when, in 1809, he suggested that all living things are made of cells.

Neither Erasmus Darwin nor Lamarck meant to imply that living things consciously strive to improve. The idea of a hare saying to himself, "I think I am going to make my fur white in the winter," was just as ridiculous to Lamarck as it is to biologists today. Both Lamarck and Erasmus Darwin felt that the drive for change was an unconscious drive to adapt to changes in the environment. The changes were brought about by the use or disuse of organs. An organ that was used would improve, an organ that was not used would degenerate. These changes would, somehow, be passed on to the future generations. The "somehow" was to escape the grasp of most naturalists for another century.

Lamarck did not originate the idea of the inheritance of acquired characteristics. This idea was commonly accepted at the time, and had been since before the time of Aristotle. Many who followed Lamarck, including the grandson of Erasmus Darwin, proposed the inheritance of acquired characteristics. Yet, year after year, in classrooms where life science is taught under a name that Lamarck proposed, he is the object of ridicule, abuse, and laughter, as the advocate of an outlandish theory.

The usual example cited to illustrate Lamarck's ideas is the evolution of the giraffe. A hoofed animal in a region where ground vegetation has become scanty strives to reach the branches of trees. Driven by hunger, it stretches its neck in an effort to eat the leaves. The neck is thus made longer. Offspring of the animal inherit the acquired longer neck. More neck stretching occurs in succeeding generations. Over a long period of time, all of this neck stretching and the inheriting of stretched necks results in an animal that can be described as a giraffe. Lamarck and other proponents of evolution were still struggling with an inadequate concept of the age of the earth when this theory was proposed.

Right or wrong, Lamarck had the courage to say what he believed. He expanded on the vague references of Buffon and solidly proposed a theory of evolution. In so doing, he focused attention on a number of biological questions that needed answering. Were species fixed or did they change? If species changed, how did they change? How were these changes transmitted from one generation to the next? Was there a basic relationship among all living things, including man? Was there some unifying principle, something that all living things had in common? If so, what was this principle? And, perhaps most important of all to many people, did all living things, including man, descend from a common origin?

These tantalizing questions were posed throughout the writings of the gentlemen scientists of the Enlightenment and of those who followed. The speculations and observations of hundreds of men were scattered in as many volumes. As the nineteenth century began, no one had offered any real scientific evidence to uphold or refute the evolution of species. Someone was needed who could gather and synthesize all this information into a plausible theory. The nineteenth century produced that someone, and the world was never quite the same after he had his say.

2.   ENTER THE MAZE ON A SAILING SHIP

Charles Darwin began life in much more fortunate circumstances than did Abraham Lincoln—also born on that same day of February 12, 1809. Both men were to be at the center of violent and bitter controversy. One man was to be cut down before he could finish his work. The other was to end a long life in bitter disappointment because he failed to find the essential support for the theory he spent a lifetime building.

Despite his background and ancestors, young Charles Darwin showed little promise in his early years; his father, a prominent and successful physician, was afraid that his son would not be able to support himself when he grew up. There was nothing in this dreamy, rather lazy boy that indicated to any observer that here stood a giant of human thought. As a schoolboy, Charles did not apply himself to such classical studies as Greek and Latin. He preferred hiking through the woods, collecting rocks, insects, and birds' eggs. His father was furious at Charles for spending so much of his time in activities that did not prepare him to make a living. When Charles was sixteen, his exasperated father reproached him by saying that he "cared for nothing but shooting, dogs, and rat catching," and that he would be a disgrace to his family.

After many urgent family conferences, it was decided that Charles would go to medical school. This was not what Charles wanted to do, but if nothing else, Charles was always the obedient son and he went off to the University of Edinburgh in the company of his brother. Charles Darwin's medical education ended when he threw up at the sight of an operation. He ran from the operating room and never returned to medical studies. This episode convinced his father that a medical career was not for Charles Darwin. Dr. Darwin did not give up easily, and he next determined that Charles would be a minister in the Church of England. Again, Charles, obedient and eager to please, enrolled at Cambridge to study theology. At long last, his father was pleased. Charles earned passing grades and received his degree in 1831. The father could say with pride that he had guided Charles into a respectable profession. He, not to mention the world, was due for a rude shock.

As Linnaeus had read botany books when he was supposed to be studying medicine, so did Darwin read books which were not in the official assignments of his work at Cambridge. He read Lamarck and the books of his grandfather. He had certainly read Erasmus Darwin while in his father's house, but apparently he did not realize the significance of his grandfather's work until the later readings in his college years. He was fascinated by Alexander Humboldt's *Travels*, an account of voyages to tropical islands. Beetle collecting, which had started as a boy's play, developed into a serious interest.

He was attracted to J. P. Henslow, a botany professor at Cambridge. He spent so much time walking the lanes of the campus with Henslow that he became known as "the man who walks with Henslow." In the course of these walking conversations with Henslow, he absorbed a great deal of knowledge in zoology, botany, and geology.

During his student years at Cambridge, Darwin had every

reason to believe that upon graduation he would enter the Church of England as a minister. However, he continued to pursue his interest in natural science. He did not know that events on a ship halfway around the world from Cambridge were going to change the direction of his life.

From 1826 to 1831, when Darwin was bouncing from one school to another, a ship called the H.M.S. *Beagle* was bouncing about the waters of South America on oceanographic missions for the British Navy. The commander of the *Beagle* was a young man named Robert Fitzroy. Besides being a naval officer of the highest ability, he was filled with a sense of duty that caused him to devote his life to the service of his fellow-men. In his self-appointed mission, he was rather pompous, overbearing, and had no use for opinions other than his own. His many contributions to naval science have been overlooked because of his association with Darwin. For example, he began the use of the word "port" to designate the left side of a ship, rather than "larboard." He felt that larboard sounded too much like "starboard," the word for the right side of a ship. The term port is still in use, and the system of storm warnings he devised is still used all over the world.

While exploring the Straits of Magellan and the southern tip of South America called Tierra del Fuego (land of fire), he was impressed with the geology of the region. Fitzroy knew very little about geology, but he felt that useful metals could be mined in the region. In the ship's log he wrote that on the next voyage of the *Beagle* he would bring "a person qualified to examine the land."

During that first voyage of the *Beagle,* a rather curious thing occurred. Some of the natives of Tierra del Fuego had stolen one of the *Beagle*'s boats. To punish the natives he took two of them captive on board the *Beagle.* For unknown reasons two more Fuegians came on board. Fitzroy was afraid that if he let the Fuegians go, they might be killed by other natives, since they were no longer in their home area. He then decided to take the Fuegians back to England, educate them, and have them introduced to Christianity. All of this he did at his own expense.

Fitzroy even gave them names: Jemmy Button, York Minister, and Fuegia Basket. The fourth Fuegian died in England.

While the three Fuegians were being educated, Fitzroy was busy with plans for the next voyage of the *Beagle*. The Fuegians were quite the hit in London, and were even received by the King and Queen. Fitzroy had been planning to send the Fuegians back again, at his own expense, to be missionaries. He was very pleased when he was commissioned to take the *Beagle* on another voyage. This voyage was to be an around-the-world survey with emphasis on South America and some of the Pacific Islands. He promptly requested of the Admiralty (British Naval Authority) that he be allowed to take a naturalist, his "person qualified to examine the land."

A Captain Beaufort was head of the Admiralty department concerned with activities such as the *Beagle*'s voyage. Upon receiving Fitzroy's request, he relayed it to a Professor Peacock at Cambridge. Peacock asked Henslow if he would like to go. Henslow said *he* could not, but that he would ask around. The post of Fitzroy's naturalist was offered to at least one more person before Henslow thought of the young man who had spent so much time walking with him.

When Darwin received the news on August 28, 1831, his immediate reaction was that his father would never let him go. Darwin was right on that score. His father felt he had bounced around enough and that it was now time to settle down to the respectable business of being a clergyman. Darwin had to write a letter of refusal to Henslow. Fortunately, Josiah Wedgwood, a man of great persuasive power, heard of Darwin's opportunity.

Josiah Wedgwood was known as a maker of a particularly beautiful kind of neoclassical pottery. He was also Charles Darwin's uncle, and the legacy he has left as the man who persuaded Dr. Darwin to allow Charles to go on the voyage will be longer-lasting than his pottery. Wedgwood argued that natural history was a perfectly respectable activity for a clergyman. Amazingly enough, Dr. Darwin was convinced and gave Charles his consent.

After Darwin's first refusal, the post had been offered to a Mr. Chester. It appeared that Darwin had missed his opportunity.

Chester refused the post, but there were other problems. Upon meeting Darwin, Fitzroy found that he did not like the shape of Darwin's nose. He thought that it was too small. Fitzroy believed that you could tell a great deal about a person from his facial features. He thought that Darwin's small nose was a sign of laziness. (Fitzroy himself had a very imposing nose.) Fitzroy took several days to overcome his objections to Darwin's nose and then offered him the post of naturalist aboard the *Beagle*. The post was without pay and would take five years of Darwin's life.

On December 27, 1831, the *Beagle,* with Darwin and the three Fuegians on board, left Plymouth, England. Darwin was immediately seasick, and remained so for quite some time. The seasickness was only the first of the many illnesses Darwin was to suffer both during and after the voyage. Just before he left, Henslow gave him a copy of a new book, Lyell's *Principles of Geology*. Henslow advised him to read it, but not to believe it. Darwin took the first part of his old friend's advice.

As Darwin sailed across the Atlantic Ocean, he still believed in two things. The first was that species were immutable, and the second was that when he got back to England he was to be a minister in the Church of England. In a letter to his sister he said that often he had a mental picture of living in a country parsonage. The observations that were to change the first belief would push the second further and further away from reality.

Darwin saw his first tropical island when the *Beagle* landed at the Cape Verde Islands off the western coast of Africa. It was on one of these islands that Darwin did his first climbing. He was to do a great deal more of it before the voyage was over. He climbed to investigate a band of limestone in the face of a cliff. In this limestone, about fifty feet from the level of the sea, he found sea shells embedded. They were the same kind of shells he had found on the beach below. According to what he had read in Lyell, some force had lifted this band of limestone that had once been under water.

He was expert at specimen collecting. After all, he had been collecting specimens since he was a boy. But now, he was collecting as a part of his official duty, which was quite a different matter from the secret hikes in the woods that used to anger his father and his teachers. During the years of the voyage, Darwin climbed mountains, collected fossils, specimens, and carefully kept notebooks of his observations.

The theory which was to change the course of history was very slow in being formulated in Darwin's mind. He was to examine his specimens and notes for many years after the voyage before writing of the ideas his observations brought about. In Brazil he found fossils of extinct giant sloths and armadillos. (He noted that, except for size, the extinct forms were very much like the living animals.) In Argentina he found the fossil bones of a large mammal called "Toxodon." In a nearby area he found some teeth of the Toxodon, the bones of an elephantlike mammal called a mastodon, and the tooth of a horse. Darwin was amazed to find the horse tooth in the same level with the remains of the Toxodon and mastodon. There had been no horses in the Western Hemisphere when the Spaniards first came. Finding the tooth together with the other fossils indicated to Darwin that horses had once existed in South America and had become extinct long before the Spaniards reintroduced them.

When the *Beagle* reached Tierra del Fuego, the three Fuegians and a missionary debarked. But when the *Beagle* later returned to this region, the Fuegians had reverted to their savage ways. The life of the missionary had been threatened by the hungry, cannibalistic Fuegians and he wisely chose to come back on board the *Beagle*. Jemmy Button, who had been so fastidious about his clothes in England, was naked, dirty, and generally unkempt. He remembered Fitzroy and Darwin and the English language, but he had no desire to return to England. Although he had returned to the primitive ways of his people he was still a marked contrast to his tribesmen. Darwin would always remember him as an example of how a man could be improved. Fitzroy was, of course, very disappointed that Jemmy would not be a missionary.

Most of the observations that led Darwin to formulate his theory of evolution were made on the Galapagos Islands. The Galapagos Islands are a group of volcanic islands about 600 miles off the coast of Ecuador. The *Beagle* landed there in

September of 1835. At first, the islands appeared rather desolate to Darwin, but he soon found that this was deceiving. Most of the islands were teeming with life. There were tens of thousands of lizardlike iguanas and many varieties of birds. One species of iguana lived on the land and another went into the ocean to eat seaweed. The latter species is the only known seagoing lizard.

The most spectacular inhabitants of the islands were giant tortoises that weighed about 200 pounds. Darwin had never seen anything like them before. As it turned out, they exist nowhere else in the world. A casual remark by the governor of the islands started a process of thought in Darwin's mind that helped him formulate his theory. The governor said that he was able to tell what island a tortoise came from by looking at it. Darwin wondered why animals living so close to each other should differ so much.

Darwin noticed that the birds on the Galapagos Islands were very much like the birds on the South American mainland. Yet the birds on a particular island were not exactly like the birds on any of the other islands. The finches were most interesting. The birds found on the various islands resembled each other, but the finches on any particular island were distinctive enough to be a separate species, differing particularly in the size and shape of beak. The beaks seemed adapted to the food that the bird usually ate. Some had long, thin beaks which were used for getting insects out of trees. Others had short, heavy beaks for cracking seeds. One species of finch was observed to take a twig in its beak and poke the stick into holes in tree branches in a search for insects.

Commenting on the diversity of birds, Darwin wrote in his notebook that such facts might "undermine the stability of species." He still believed in the immutability of species, but he was beginning to ask himself questions. He wrote that if species were fixed, it was very difficult to explain the situation in the Galapagos Islands. These islands were only a few miles apart and they all had the same physical conditions. Yet, each one had its own distinctive, but similar, species. If each species had been created on the particular island, then why were they so similar? Why did a tortoise on one island have an oblong, grayish shell and a tortoise on another have a rounder, black shell? He remembered that the birds on the Cape Verde Islands off the coast of Africa had resembled the birds of the African continent. Why were the extinct giant armadillos so much like living armadillos? He asked the same questions about the fossil remains of sloths which were so much like the sloths still existing in South America.

The *Beagle* continued on its voyage for another year after its stop at the Galapagos. Darwin continued to collect, observe, and keep careful records in his notebooks. He also made extensive geological observations. His theories of the formation of coral reefs are still regarded as essentially correct.

The *Beagle* returned to England on October 2, 1836. Attempting to establish order out of the mass of similarities and differences he had seen in the past five years was to occupy the rest of Darwin's life. Although he lived to be seventy-three it was not long enough. The life of no one man could have been long enough to understand all of what Darwin had seen.

Darwin found an apartment in London and set to work on his collection of specimens. The vision of the country parsonage now was gone forever. He wrote all of his thoughts and observations in a series of notebooks. In July of 1837 Darwin started a notebook entitled "Transmutation of Species." In this notebook he started to build what he called "my theory." Using his observations in South America and the Galapagos as evidence, he proposed that species descend with modifications from a common ancestor. He thought of the descent not as a straight line, but as a series of branches, somewhat like a tree. He made a rough sketch of an "evolutionary tree" in his notebook. Darwin was certainly not the first person to propose such a theory of evolution, but no one before him had come prepared with such a massive arsenal of evidence as he had accumulated during his travels on the *Beagle*.

He worked slowly, piecing his evidence together with the methodical precision of a watchmaker. It was not until 1842 that he put together even a rough draft of a paper to be submitted for publication in a journal. He was not exactly idle between 1837 and 1842. He married his cousin, Emma Wedgwood, in 1838. His first book, *The Journal of Researches* was published in 1839.

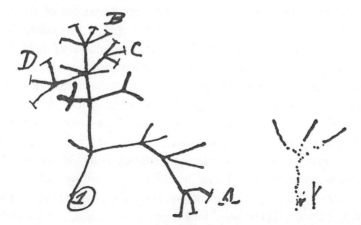

This book, known as *The Voyage of the Beagle* in the United States, was an account of his geological and biological observations during the voyage. He did not propose a theory of evolution in his first book. By 1842 he had two children and had bought a house. It was in this house, called Down House, about sixteen miles from London that he spent the rest of his life.

In January of 1844, Darwin wrote a letter to his friend, Sir Joseph Hooker, in which he stated that "I am almost convinced that species are not immutable." Darwin had indirectly helped Hooker to get an appointment as a naturalist for an expedition to the Antarctic. They became close friends, so close that it was Hooker to whom Darwin usually expressed his ideas before he published them. Darwin had also developed a friendship with Charles Lyell, the geologist.

Very soon after his letter to Hooker, Darwin began to work on his *Essay on Species*. He had previously been very much impressed by Thomas Malthus' *Essay on Population* written in 1793. Malthus had expressed concern over the rapidly increasing human population. He expressed the view that human population would outgrow the food supply if not held in check by such things as war, famine, and disease. Malthus spoke of a consequent struggle for existence. Malthus' work provided Darwin with a theme about which to build his theory of natural selection.

Darwin completed the *Essay on Species* in July of 1844. In the *Essay*, Darwin stated that organisms are able to reproduce more of their kind than can be supported by the resources of the land. Since there was no evidence that any one species was overrunning the world, it could be assumed that large numbers of individuals died, or were killed, before they reached maturity. Following this premise, Darwin presented his main argument. Those organisms which did survive to maturity did so because they were better equipped to obtain the necessities of life. Less-favored organisms lost out in the struggle to obtain food, water, and other necessities.

Why were some individuals in a species better equipped than others? Darwin answered this question by pointing out that individuals within a species are not exactly the same. That is, some show variations. These variations happen by chance. An individual with a particular variation will pass this variation on to its offspring. The variation may be something which gives the individual a competitive edge in the struggle for the necessities of life. Such favored individuals are more likely to live long enough to reproduce offspring which will inherit the favorable variations. Darwin pointed out that it was a demonstrated fact that offspring tend to resemble their parents.

He concluded his arguments by stating that each succeeding generation would maintain and improve favorable variations. Generations of variations would produce new varieties that would eventually become new species. He noted that the development of new species was more likely to happen if the varieties were isolated from each other.

Darwin did not limit himself to physical variations. He also discussed the evolution of instincts which increase the organism's chances of survival. His theory can be summed up in a statement he made a few years before he wrote the *Essay;* "favorable variations would tend to be preserved and unfavorable ones to be destroyed."

Darwin did not publish his *Essay.* As far as is known, only Hooker was allowed to read it. Shortly after he wrote the *Essay* he turned his attention to the study and the classification of barnacles. Hooker and Lyell urged Darwin to publish a paper on his theory as soon as possible as someone else might do so before. Darwin replied by saying that his barnacle project was important to him. He expected the work to improve his understanding of the problems of designating species. He also felt that he had to make some tangible contribution to knowledge of biological classification.

Barnacles are related to crabs, lobsters, and shrimp. They are familiar to boat owners as the growths which have to be scraped from the bottoms of boats from time to time. Darwin was interested in them because they showed much variation within the group and previous attempts at classification had been inconclusive. Darwin spent some eight years working on barnacles. During this time he published another book, *The Geology of South America*, and his wife had three more children. In all, he worked with about 10,000 specimens of fossil and living barnacles. He finished in 1845, and again turned his attention to his work on species. His friends, Hooker and Lyell, felt that it was high time.

Darwin was still in no hurry. He was very cautious and wanted to gather more evidence. He carried out breeding experiments with animals and joined a pigeon-fancier's club. One of his strongest arguments for variation and natural selection was that man had been successfully breeding variants of pigeons, some of which were quite bizarre. Darwin believed that all the varieties had come from the original blue-gray rock pigeon, the common variety seen in parks and on city streets. On several occasions he crossed fancy varieties of pigeons and obtained offspring that reverted toward the common rock-pigeon-type. From these experiments, Darwin surmised that the characteristics of the rock pigeon were "hidden" in the other varieties, but he did not say how this occurred or why the hidden characteristics should emerge after many generations.

Darwin had stated that newly formed islands could become inhabited when seeds and eggs were carried to them in ocean currents. His friend, Hooker, disagreed with him, as did Lyell. Hooker believed that islands were once part of a large land mass which was broken up into islands by some geological catastrophe and that living things became isolated on them. Hooker did not believe that seeds could survive in salt water. To prove his point, Darwin soaked seeds in sea water for different periods of time.

When he planted the seeds he found that some of the seeds grew into plants after being in salt water for over 130 days. Calculating from the known rates of ocean currents, Darwin told Hooker that the seeds could be immersed long enough to float from the West Indies to Scandinavia. Hooker grudgingly admitted defeat. Darwin's children had thought these experiments great fun and were delighted when the seeds "beat Dr. Hooker." Darwin also thought seeds could be transported in the bodies of animals and by icebergs. He found evidence to support these ideas.

By 1856 he felt he had enough evidence to begin a large work on species. He decided to include the words "Natural Selection" in the title of the book. It took him two years to write ten chapters.

His orderly progress was interrupted on June 14, 1858, when he received a letter from Alfred Russel Wallace. Wallace was a naturalist and a friend of Darwin who had been doing some work in Southeast Asia. The letter was a short paper from Wallace in which he stated a theory which was almost exactly like Darwin's own theory of evolution by natural selection. Darwin sent the letter to Lyell and admitted that Lyell's "I told you so" was correct.

Wallace had known nothing of Darwin's writings yet his work was so much like Darwin's that even some of the section headings were the same. Darwin felt that all his many years of work were for nothing since he could no longer claim originality. He felt that it would not be honorable for him to publish his work even though he had started before Wallace.

Lyell and Hooker came to Darwin's aid. They pointed out that Darwin's letter to Hooker in 1844 and a letter to an American botanist, Asa Gray, in 1857, proved that Darwin's work was original and had not been stolen from Wallace. Hooker and Lyell proposed that Darwin and Wallace should present a paper together, but they insisted that Darwin's letters to Hooker and Gray be included in the paper.

On July 1, 1858, the joint paper of Darwin and Wallace entitled *On the Tendency of Species to Form Varieties; and on the Perpetuation of Varieties and Species by Natural Means of Selection* was read before the Linnaean Society. Darwin did not attend the meeting. The paper was presented by Hooker and Lyell.

The contents of the paper were basically the same as the *Essay on Species*. Darwin added some comments on sexual selection. He said that features such as the bright feathers of the peacock had evolved because the bright feathers attracted the peahen. Peacocks with bright feathers were more successful in getting mates. Therefore the characteristic was maintained.

The members of the society listened politely. Whether they liked it was hard to say. No one there offered any comments or questions. Darwin was a little disappointed at the lack of response from the members of the Linnaean Society. He felt that his book on natural selection would be received with greater interest. And fifteen months later, when his book was published, he was more correct than he had hoped to be.

Darwin's publisher felt that *An Abstract of an Essay on the Origin of Species and Varieties Through Natural Selection* was too long for a book title. The publisher suggested *The Origin of Species* and Darwin reluctantly agreed to the change. The publisher did not think the book would be very successful, but on publication day, November 24, 1859, the first printing of 1250 copies was completely sold. This, of course, pleased the publisher who was afraid he had printed too many copies. He happily ordered another printing.

The response to the book was immediate and stormy. Darwin, of course, expected to be challenged by botanists, zoologists, etc. But, although he expected some public reaction, he was not prepared for the storm of violent protest that greeted his book. This storm continues even today.

Sides were quickly drawn up in the scientific world. Hooker and Lyell quickly rallied to Darwin's defense. They were joined by Thomas Henry Huxley, a well-known English biologist. Most of the public outrage centered about the last sentence of the last chapter. "Much light will be thrown on the origin of man and his history," it read. Darwin had avoided any discussion of man in *Origin*. He knew that the inclusion of man in the book would bring him a lot of trouble. But he felt that he would not be honest

with himself if he made no mention of it at all. The last sentence
was a compromise.

Darwin was accused of saying that men came from monkeys.
Darwin never made such a statement, even in his later book, *The
Descent of Man*. But even today there still are people who believe
that Darwin's major contribution was a theory that man came from
apes and monkeys.

The misunderstanding of Darwin's thoughts on the origin of
man were evident at a meeting of the British Association for the
Advancement of Science held in June 1860. At this meeting, a
Samuel Wilberforce delivered an attack on Darwin's book. Wilberforce was not at all qualified to enter into any serious discussion
of biological matters, but he had been coached by Richard Owen,
an anatomist and paleontologist, who still held to the views of
Cuvier. Darwin was not at the meeting, but Huxley and Darwin's
ex-shipmate, Fitzroy, were there—on opposite sides.

In this talk, Wilberforce said there was nothing to the idea of
evolution. He said that rock pigeons were rock pigeons, and would
always be rock pigeons. He then turned to Huxley and asked if it
was through his grandmother or grandfather that Huxley claimed
descent from a monkey. Huxley leaped to his feet and replied in
words to the effect that he would rather have an ape for an ancestor
than a brilliant man who entered into discussions of scientific
matters of which he knew nothing. The audience reacted wildly
to this obvious insult to Wilberforce, who was a bishop in the
Church of England. A certain Lady Brewster fainted and Captain
Fitzroy stalked about the room holding a Bible over his head and
crying out, "The Book! the Book!" Much to Darwin's distress,
natural selection had become a religious issue.

Upset as he was at attacks from the general public on religious
grounds, Darwin was much more concerned with attacks from
other scientists. The Wilberforces and others like him were angry
with Darwin, only because he was a comfort destroyer. Their
egos were hurt. Their protests were only the cries of frightened
people who could not face the possibility that they, as men, might
have evolved just like any other animal. All of their screaming
could not do very much damage to his theory. Attacks from other

scientists based on available scientific evidence were another matter. And Darwin found it difficult to come up with answers.

Lord Kelvin, the famous English physicist, offered evidence that
the earth was not old enough for evolution to have taken place.
Darwin, being no physicist, could not offer effective counterargu-
ments. Kelvin was wrong, but this was not to be known until
another century.

Darwin's critics wanted to know just how these variations came
about. Why should there be any variations at all? Living things
seemed to get along with what they had. Even if there were
variations, how were these variations inherited by offspring? To
questions such as these, Darwin had no immediate answers.

Some critics maintained that even if variations did occur they
could not be perpetuated in future generations. Since the "un-
varied" form was the more numerous in a species, the variation
would be canceled out. Most people at the time, including Darwin,
believed in blending inheritance: that is, that the characteristics
of the parents tend to blend in the offspring. For example, if an
animal with black fur mated with an animal with white fur the
offspring would tend to be gray. Following the blending idea,
the critics said that any variations would be "blended out" or
"swamped" by the more prevailing characteristics. The only way
in which chance variations could effect a permanent change in a
species was if the same variation occurred in a male and a female
and these two mated. Even then, the critics said, the variation in the
offspring would still tend to disappear.

Darwin's critics may have had plenty of questions, but they
did not have any more answers than did Darwin. During Darwin's
time there was very little information available to biologists which
would either support or condemn Darwin's theory. The role of the
male sperm cell in reproduction was not clear. Nothing at all
was known about how characteristics are inherited from parents.
There were still many who believed that fully formed animals
could spring forth from nonliving materials such as mud. The
lack of knowledge about reproduction and inheritance made Darwin's theory very difficult for many to accept. The emotional public outcry against the theory certainly influenced some biologists

who otherwise may have examined the theory in a more un-prejudiced way.

Darwin also had problems with some of his supporters who prodded him into making statements he could not support. A certain Samuel Hearn observed some bears swimming with their mouths open. This led Hearn to speculate that whales evolved from bears. Darwin was persuaded to include this idea in one of the revisions of *Origin*. His critics attacked the idea as completely ridiculous. The story circulated that Darwin had claimed a bear could turn into a whale. Darwin cut the statement out of the next edition of his book. (As it turns out, Hearn may not have been completely wrong).

Ernst Haeckel, a German embryologist, was an ardent supporter of Darwin. Haeckel proposed that in their embryonic development, organisms went through stages which represented their adult evolutionary ancestors. This idea, called the theory of recapitulation, was incorporated into some of the later editions of *Origin*. It was an intriguing idea that held on for a long time. Haeckel was also inspired to make beautifully drawn charts which were supposed to show the stages of man's evolution from the amoeba. Haeckel included pictures of "animals" which never existed. Haeckel's fanciful charts were so ridiculous that they influenced some biologists not to support Darwin.

Herbert Spencer, another of Darwin's friends, suggested that "survival of the fittest" was better than natural selection as a title for Darwin's theory. Darwin accepted survival of the fittest, much to the disgust of Huxley. It was a very poor term since it gave a "tooth and claw" idea of evolution. The term has resulted in a lot of misunderstanding of Darwin's theory. But it made a good heading for newspaper stories, and it persisted.

Darwin never stopped working to obtain more evidence. He carried out a series of plant-breeding experiments. He worked with a variety of snapdragons which usually have irregular flowers. He crossed the irregular variety with a variety that has regular flowers. All the plants obtained from the cross-pollination had irregular flowers. When these irregular flowers were allowed to pollinate themselves, the resulting flowers were regular and irregular. He obtained thirty-seven regular and eighty-eight irregular. But he apparently failed to see any significance in these figures.

He may not have remembered what he wrote to Huxley in 1858 when he was preparing the first edition of *Origin of Species*. ". . . I have lately been enclined to speculate . . . that propagation by true fertilization will turn out to be a mixture and not true fusion of two distinct individuals, as each parent has its own parents and ancestors. I can understand no other view in which crossed forms go back to such an extent to ancestral forms."

Darwin observed that plants which are self-pollinated tend to produce generations of offspring with the same characteristics, while those which are cross-pollinated have offspring with a variety of characteristics. In later years, he made observations of plants which have adaptations to prevent self-pollination. He felt that cross-pollination was beneficial to the plant in producing new varieties. This led him to comment that the offspring of the first generation tend to assume the characteristics of one of the parents, while the second generation shows characteristics of both parents. He wrote that he knew of some varieties which did not seem to blend but which had characteristics that were transmitted intact from one generation to the next.

Darwin had to answer the critics. The source and transmission of variations was the weakest point of his theory. For his answer, Darwin went back to ideas which had been put forth by the ancient Greeks and, more recently, by Lamarck. Apparently he drew very little from his plant-breeding experiments in formulating this theory which he called "pangenesis." He proposed this theory in his book, *Variation of Plants and Animals Under Domestication*, published in 1868. In this theory he stated that when organs or parts of organisms are used, the organs produce particles called "gemmules." These gemmules are carried in the blood stream of the animal to the reproductive organs where they are incorporated into the reproductive cells.

Darwin was very careful with his use of the word "cell." The idea that all living things are made of cells was not completely accepted and he wanted to cover himself against his critics as much as possible. He said that gemmules were gathered from all parts of the body rather than from cells. The gemmules passed on to future offspring would result in future generations inheriting favorable variations from the parents. He went on to say that the gemmules were distributed to various parts of the body where they caused the development of various organs. For example, a wading bird might find it necessary to go out into deeper water to get food. It would make swimming motions with its feet. These motions cause the production of gemmules which are carried to the reproductive organs, where they are incorporated into sperm or eggs. Birds which develop from the sperm or eggs containing the gemmules from the feet of the parents will have feet better adapted for swimming, according to the theory of pangenesis. The degree of development of the feet is dependent on the number and quality of gemmules.

He explained such things as ordinary rock pigeons resulting from the mating of two varied types by saying that there was a "reserve of gemmules." This reserve could, under certain conditions, determine the characteristics. According to this theory, when an organism used a particular part in response to the environment, this part produced gemmules. If the part was not used, it did not produce gemmules, or at least not as many as a part that was used.

The theory of pangenesis was essentially Lamarck's idea of the inheritance of characteristics acquired by the use or disuse of a part. Darwin expanded on it with the gemmule idea. It was an attempt to save natural selection by drawing on the little that was known about reproduction, which was very little. At the time Darwin proposed pangenesis he thought that several pollen grains or sperm were necessary to fertilize an egg. Very little was known about development from fertilization to birth (or hatching).

Pangenesis was vigorously attacked even by many who supported Darwin in natural selection. The theory was effectively knocked down by Darwin's nephew, Francis Galton, who was trying to provide evidence to uphold the hypothesis of his uncle. Galton reasoned that gemmules must always be in the blood. He transferred the blood from a rabbit of one variety to a rabbit of another variety. He expected that the offspring of the second rabbit would inherit the characteristics of the first rabbit. Amazingly, the rabbits which received the transfused blood lived, and lived long enough to have baby rabbits. The baby rabbits were like their parents in every case and showed no signs of having inherited the characteristics of the blood donor. On hearing of the results, Darwin made a strategic retreat. He said that gemmules were not necessarily in the blood. He pointed out that plants, which have no blood, have gemmules. But again, Darwin's supporters had done more damage than the critics. Natural selection was in trouble.

Darwin was discouraged by the lack of success of pangenesis. He left the problem of species in order to finish a book which everyone had expected him to do. This was *The Descent of Man,* published in 1871. His arguments—that man descended from other species just like all other organisms—were very well stated in the book. It created a sensation and sold very well. Again he lacked any real evidence. The finding of the remains of fossil men and premen was to come some years in the future.

In the eleven years between the publishing of *The Descent of Man* and his death in 1882, Darwin wrote ten more books. He received many honors, but he was saddened, not only at his inability to answer his critics, but at the bickering of his supporters. They were arguing over many fine points, not knowing that they all really were in basic agreement.

His health deteriorated. He had been ill for much of his life after the voyage of the *Beagle.* The doctors could not diagnose his illness, and many thought his problems were brought on by mental tensions. He probably had picked up a tropical disease in Brazil. It was known that he was bitten by insects of the Triatomid family. This insect is now known to transmit a disease called Chagas' disease. The cause of the disease and the fact that the Triatomids transmit it were not known until many years after Darwin's death.

That the cause and nature of his disease were unknown until after his death is an appropriate commentary on his life. Almost everything that Darwin needed to know to give natural selection solid support was not known until after his death. An unknown priest working in a garden in Austria and hundreds of men looking down into microscopes were to provide the evidence that would save natural selection.

The ironic thing is that much of the evidence was there when Darwin was alive, but everybody was too busy to notice.

3. THE ANSWER IN THE GARDEN

When Darwin published his ideas on pangenesis in 1868, he probably did not even know of the existence of an obscure journal called the *Journal of the Brunn Society for the Study of Natural Science*. He certainly did not know of a paper that had been published in this small journal some two years earlier, entitled, "Experiments in Plant Hybridization." This paper had been written by an Augustinian monk named Gregor Mendel. Even if Darwin had taken time from his busy schedule to read an amateur's paper in a small, third-rate journal, he probably would have failed to grasp its significance. This is a reasonable assumption. No one else in Darwin's time understood its significance either. No one, that is, except Gregor Mendel.

If the name Gregor Mendel had been mentioned to any of the eminent natural scientists of Darwin's day, the reply would almost certainly have been, "Who is Gregor Mendel?" Even the great botanist, Carl Naegli, to whom Mendel wrote of his experiments, would probably have had some difficulty recalling the name.

If Darwin had read Mendel's paper and understood it, he would have been rather surprised for a number of reasons. For one thing, Darwin had done the same experiments that Mendel reported in his paper. Darwin had even obtained some of the same results, but the significance of these results escaped Darwin. Darwin was certainly not stupid. However, he was rather weak in mathematics, and mathematics was the tool that enabled the obscure Gregor Mendel to discover a few basic principles that gave substance to the theory of evolution.

The weak point in the theories of Darwin and Wallace was the lack of knowledge of the mechanism of heredity. No one knew much about inheritance in those days. There were a lot of ideas, but ideas were nothing more than speculation. None of the ideas of inheritance were backed up by any experimental evidence.

Most of the ideas on heredity held that characteristics of the parents were somehow blended in the children. One of the most prevalent ideas was that inheritance was determined by something in the blood of the parents. Even today, we hear expressions such as "full-blooded Indian" and the like. Of course, it was a rather difficult matter to explain just how the "something" in the blood got into the sperm and eggs.

Darwin spoke of gemmules in the blood in his pangenesis theory. The pangenesis theory did not have a very long stay in the minds of men, but there are still many people in the world who blithely utter such expressions as "blood will tell," without realizing they are stating a belief in the blood theory of heredity.

That very little was known about heredity in Darwin's time is not surprising. Not too much was known about the reproductive processes of living things. Although Leeuwenhoek had observed sperm cells in 1677, it was not until 1876 that sperm cells were proven to be essential agents in fertilization. Many people still believed in spontaneous generation and stoutly defended such ideas that earthworms sprang forth fully formed from the soil without benefit of parents. An even more ludicrous idea still had a few followers. This idea—that a human baby, for example, was fully formed in the sperm or egg, and only expanded in size during the gestation period—was widely held in the seventeenth and eighteenth centuries. One of the most violent controversies in the history of biology was waged in the early eighteenth century between the "spermists" and the "ovists." The spermists held that the preformed baby or "homunculus" was in the sperm; the ovists argued that it was in the ovum, or egg.

Just what did people know about heredity in the early nineteenth century? Not very much. It was known that organisms produced more organisms like themselves. Dogs reproduced dogs, snakes reproduced snakes, milkweed reproduced milkweed. With the discovery of the basic facts of fertilization, it was evident that "dogness," "snakeness," etc., were in the sperm and eggs of the parent organisms. Just what was in the sperm and eggs that determined that offspring were going to be like their parents was unknown.

That like reproduces like is obvious. Even prehistoric men knew this. However, there was much that defied explanation. It was observed that dogs always produced dogs, but that no two dogs were ever exactly alike. A human child may look like its parents, but it never looks exactly like its parents. Brothers and sisters tend to look alike, but except in the cases of identical twins, they never look *exactly* alike. Perhaps most disturbing to investigators was that there was some kind of pattern to the way things were inherited. But attempts to discover and define this pattern were frustrating.

There was a growing interest in plant and animal breeding in the nineteenth century and, therefore, a need to know more about the nature of inheritance. It was logically assumed that selecting animals with favorable characteristics for breeding stock would result in young with the same favorable characteristics. Plant and animal breeders were disconcerted to find that the results of the matings were not always as logical as the selection of the parents. The same is true today, but at least we know why.

It was observed that a particular characteristic present in the parent might not show up in the offspring. The same characteristic might reappear several generations later, even though it had remained "lost" in the intervening generations. Other characteristics were observed always to breed true. That is, they would reappear generation after generation. Breeding experiments bogged down in attempts to find a pattern in the seemingly infinite combinations of characteristics.

Despite these difficulties, animal and plant breeders of the time had developed many breeds and varieties of organisms. A dachshund, for example, is a breed of dog. Essentially this means that dachshunds possess a set of true breeding characteristics that will always appear in the offspring if you mate a dachshund with another dachshund. When this situation exists in a plant, it is referred to as a variety. The term purebred is also used to describe a breed or variety. By the nineteenth century it was possible for a farmer to buy seed that was guaranteed to be purebred for certain characteristics. No one knew why the seed was purebred, but if the seed was sold with a money back guarantee, the farmer felt that was all he needed to know.

Even before the nineteenth century, breeders began to become interested in the production of *hybrids*. A hybrid is the result of mating two different purebreds or varieties (or species). Frequently the result was an organism that combined the best characteristics of the two breeds. As might be expected, the opposite was frequently the case. Hybrids of wheat that were resistant to disease and had good per acre yields were developed. Many people took up the breeding of hybrids as a hobby.

One such hobbyist was a teacher in a primary school in the town of Heinzdorf in Austrian Moravia (now Czechoslovakia). During the 1820's and 1830's, the teacher, Thomas Makitta, taught techniques of fruit culture and hybridization to interested boys at the school. Fruit cultivation was not part of the official curriculum at the school, but Makitta conducted this activity as a sort of club, something like the "extracurricular activity" of today.

One of Makitta's pupils, a boy named Johann Mendel, was particularly enthusiastic about all aspects of plant breeding. This boy devoted all his spare time to plant breeding and still managed to bring home good report cards. As Makitta watched Johann's progress, he came to feel that it would be tragic if the boy did not continue his education after grade school.

Young Mendel did not need any persuasion. He wanted to continue his studies, but there were many problems. Johann Mendel was the son of a peasant farmer. In the Europe of the mid-nineteenth century, it was uncommon for peasant boys to go beyond grade school. The main problem was money. Schools were not free. Most peasants just could not afford the tuition.

Makitta knew that peasants respected the advice of authority, especially if the authority was a titled nobleman. This attitude was so prevalent among peasants so as to be a tradition. Makitta influenced the wife of a local nobleman, the Countess Waldburg, to persuade Mendel's parents to provide for the further education of their son.

Mendel's parents, Anton and Rosina, saved their pennies and sent their son to *Mittelschule* (junior high school) and *Gymnasium* (high school).

In 1837, when Darwin started his Transmutation of Species Notebook, Mendel was ready to start college at the Philosophical Institute at Olmhutz. Oddly enough, this young man whose name and work were to become basic to all students of biology, never took a course in natural science until after he graduated from college. Natural science or natural history included what we today call biology. Mendel took the required classical studies, such as mathematics, literature, and philosophy.

In 1838, a tragedy threatened to end Mendel's education. As it turned out, this tragedy steered Mendel into the situation that enabled him to do the work for which he is remembered. Anton Mendel was seriously injured in an accident and could no longer work. Johann was now without funds to continue his education. He tried to continue with money earned from tutoring. But the strain was too much and he became ill. He dropped out of college for a year to recover and to try to earn some more money from tutoring.

In 1839 he returned to college with a little tutoring money and some money his sister had saved. This small sum soon ran out and Mendel was in a desperate situation. He did not have even enough money to buy food. At this point, he was seriously considering stopping his education and taking some kind of job. He sought advice from one of his teachers, Frederick Franck, an Augustinian monk. Father Franck advised Mendel to enter the Augustinian monastery in the town of Brunn. In 1843, Johann Mendel was accepted into the order. He was required to take a new name. He chose Gregor.

Mendel was indeed fortunate to enter the monastery. Some of the most outstanding scholars of the time were Augustinian Brothers. Mendel continued classical studies, but devoted most of his free time to natural science. Mendel was at the monastery fifteen years before he did the experiments that resulted in his 1865 paper. During that time he taught as a substitute teacher at a nearby *Gymnasium*.

He wanted to be a full-time, qualified teacher of natural science. He took the required qualifying examinations and failed. Mendel had never taken a course in natural science. He was entirely self-educated in this subject and was unfamiliar with much of the content of the examinations. The monastery then sent him to the University of Vienna where he studied zoology, botany, microscopy, physics, and chemistry.

Mendel did very well in his studies at the University, where one of his professors, Franz Unger, was almost expelled from the University because he had made statements on the immutability of species. Professor Unger later became one of Darwin's strongest supporters.

Mendel now felt confident enough to take the examinations again. He failed the examinations again. There is some evidence that the readers of the examinations discriminated against him because he was a Catholic priest.

Mendel was, of course, very disappointed. But again, what initially seemed a tragedy, pushed him one step closer to greatness. He could not be a full-time teacher, but he still did some substitute teaching which took up very little of his time. He found that he had enough time to devote to the study of plant breeding. There was a small plot of ground near the main monastery building that impressed Mendel as being quite suitable for a garden. Mendel obtained permission to do some gardening on this little plot. Many of the brothers in the monastery were quite amused by Brother Gregor's "puttering." Mendel went quietly about his work.

Mendel started his work by trying to produce hybrids of several varieties of flowering plants. He accomplished this by the laborious means of *crossing*. In making a cross he would carefully gather the pollen from a flower and sprinkle the pollen on the pistil (female reproductive structure) of another flower. He would then cover the flower with a bag to prevent unwanted pollen from reaching the flower. Male gametes in the pollen would fertilize the egg cells in the female parts of the flower. The result was seeds that represented the "children" of the crossed flowers which were the "parents." Mendel called each new crop of seeds a generation.

According to the ideas prevalent at the time, a new generation was supposed to have characteristics that were the result of blending of the characteristics of the parents. For example, if you crossed a red flower with a white flower you would expect the offspring to be pink. In the course of his work, Mendel noticed that some characteristics would appear unchanged generation after generation. Sometimes a particular characteristic possessed by one or both of the parent plants would disappear in the offspring. This particular characteristic might reappear unchanged in the "grandchildren" or "great-grandchildren." To Mendel these observations seemed to contradict the generally accepted idea of blending inheritance. Mendel was curious. Why did some characteristics disappear and reappear later, while others persisted in every generation and others seemed to blend? Was there a pattern to the way these characteristics were inherited or was all heredity haphazard—without rule or pattern? Mendel decided to investigate the problem further. If there was a pattern to inheritance he was going to find it.

Mendel read all the available books on plant breeding and many papers in scientific journals. Checking on other work that has been done is part of the scientific method and is as much a part of scientific investigation as the actual experimental work. It is necessary to know if anyone else has done the work you are thinking of doing. Mendel hoped that he could gain some insight into the problem he planned to investigate.

Mendel found that all previous breeding experiments had yielded chaotic results. From the papers he read, it appeared that no one had continued experiments into several generations. He also noted that most of the workers before him had tried to draw conclusions from the appearance of the entire animal or plant. The results of these crosses were a multitude of combinations of characteristics that did not seem to follow any rules. What seemed most strange to Mendel was that few of the workers had bothered to count the offspring.

Most of the people engaged in breeding experiments prior to Mendel believed in the idea of blending inheritance. Mendel was remarkably free of such preconceptions. Perhaps if he had been in the mainstream of the scientific world he would have held to the prevailing ideas of the time. In the seclusion of the monastery, he was free to determine his course of action without help or hindrance from well-meaning colleagues.

He did some experiments with white and gray mice. The results were inconclusive. He needed an organism that he could control in a more effective fashion. Mendel decided to use the edible garden pea for his experiments. This plant was suitable for a number of reasons. The garden pea is ordinarily self-fertilized. That is, in any flower the pollen produced by the stamen (male reproductive structures) fertilizes the eggs in that same flower. And the shape of the flower protects the reproductive structures from contact with unwanted pollen. By application of a few simple but painstaking techniques, Mendel could cross-pollinate to produce hybrids or allow the plants to self-pollinate.

To begin with, Mendel needed pure breeding plants. This was no great problem for him. He was a farmer's son and knew of several seed companies that sold seeds guaranteed to be purebred. He ordered thirty-four varieties of seeds. The supplier promised his money back if the seeds did not breed true. Mendel was not one to trust to guarantees. He planted the seeds and the seeds of the offspring for two years to satisfy himself that the seeds were indeed purebred. Then, in 1858, the same year that Darwin and Wallace presented their joint paper before the Linnaean Society, Mendel began the experimental phase of his work.

Mendel had plenty of time to carry out his work. He was under no pressure and had no deadlines to meet. Remembering the chaotic results of earlier experiments, he decided to study only one or a few easily compared characteristics at a time, rather than the entire plant. Without fanfare, and unknown to anyone including Mendel himself, the gentle monk was starting a revolution in the minds of men equal to that started by Darwin in the same year.

He selected seven characteristics:

1. the color of the seed—yellow or green
2. the shape of the seed—round or wrinkled
3. the color of the seed coat—white or gray
4. the form of the ripe pod—smooth or wrinkled between the seeds
5. the color of the unripe pods—green or yellow
6. the position of the flowers—distributed along the stem (axial) or bunched at the top of the stem (terminal)
7. the length of the stem—long (6 to 7 feet) or short (9 to 18 inches)

His first effort was to produce hybrids of the wrinkled seed and the round seed varieties. He planted seeds of each variety in separate areas of the garden plot. As soon as buds formed on the vines, he carefully opened each bud and removed the stamens. In so doing, he prevented self-fertilization. He tied little calico bags around the flowers to prevent pollination by unwanted pollen. He collected pollen from the "round flowers" (flowers of plants that had grown from round seeds). He sprinkled this pollen on the pistils of the "wrinkled" flowers. He reversed the pollination procedure on some plants by sprinkling wrinkled pollen on round pistils. This was done to see if the sex of the parent might influence the results. In the seed-shape experiment, he painstakingly pollinated 287 flowers on 70 plants. Now all he could do was what any other gardener has to do: take care of his plants and wait for them to grow.

During the long weeks of waiting for sun, water, and soil to do their work, Mendel must have speculated on what the results might be. According to what was believed at the time, the results should have been a mixture of round and wrinkled seeds and some seeds that appeared intermediate between the two characteristics. When the plants were mature, Mendel opened the pods. In the first pod, he found only round seeds. Again and again, he found only round seeds. When he had harvested all the seeds he found that every seed was round. The wrinkled characteristic, possessed by half the parents of these seeds had disappeared as though it had never existed! Of course, Mendel and others had seen this "disappearing" of traits before. But apparently no one, Mendel included, had ever observed such complete disappearance. In addition, there was no evidence of blending. All of the seeds were round; not one of them had the slightest appearance of being wrinkled.

Other investigators, under the influence of preconceptions, might have discarded these results as some kind of accident. Perhaps Mendel would have thought these results accidental had he not achieved similar results with the other crosses. In every experiment, he found that one characteristic disappeared in the progeny of the first cross. In the tall-short plant cross, he found that the short characteristic disappeared in the first generation. Of course, he had to plant the seeds to observe this. The characteristic that persevered in the first cross, he called the *dominant*. The one that disappeared was called the *recessive*. He called the progeny of the first cross the *first generation*.

Mendel was anxious to see what he would get in the second generation. He saved the seeds he obtained in the first cross and planted them the next spring. When the plants grew to maturity, he allowed them to fertilize themselves. This, in effect, meant that all the parents of the succeeding generation were round seed plants. Again, Mendel had to wait patiently through the summer months for the plants to grow and mature.

Upon opening the pods of the mature plants he found both round and wrinkled seeds. In many instances he found both kinds of seeds in the same pod. The "lost" trait of wrinkled seed had reappeared in seeds whose parents were all round seeds! Round and wrinkled parents produced all round offspring; all round parents produced round and wrinkled offspring! At this point, many investigators would have given up the experiment as hopeless and despairingly cried that there was no pattern to inheritance. Not so Mendel. He did something few had ever done before—he counted the offspring. Counting may seem a simple enough thing to do. Yet, this simple task of counting proved to be the measure of Mendel's success where others had failed.

Mendel gathered and counted 7234 seeds in the seed-shape experiment. Of these seeds, 5474 were round and 1850 were wrinkled. The ratio of round seeds to wrinkled seeds was very close to 3:1. He obtained similar results in his other plantings. In crossing tall and short plants, the first generation was composed of all tall plants. The second generation (progeny of all tall parents) exhibited tall and short plants in a ratio close to 3 tall to 1 short. In all the other crosses the results were similar. Mendel noted that the trait he had designated as dominant was the more numerous in the second generation.

Mendel carried his experiments through to a third generation. He planted the round and wrinkled seeds from the second generation and allowed these plants to fertilize themselves. When the plants had matured, he harvested the seeds. He found that the wrinkled seeds had all produced plants that bore only wrinkled seeds. In further experimentation, he found that generation after generation of wrinkled seeds produced only wrinkled progeny. The situation was quite different with the round seeds. Of the round seeds he planted, two thirds of them matured into plants that produced round and wrinkled seeds in a 3:1 ratio. One third of the round seeds matured into plants that produced round seeds only. Mendel had planted a lot of peas and gathered some rather intriguing data. Now he had to figure out what it all meant.

During the winter months, Mendel could not do any planting, but he was far from idle. He spent many hours at his desk trying to interpret the work he had done. Mendel did not know exactly what it was that caused his seeds to be round or wrinkled, with stems short or tall, etc. He surmised that this "something" that determined the characteristics had to be in the gametes (pollen and eggs) of the parents. For lack of a better term, he referred to the something as *merkmalen* which means "factors." Mendel surmised further that since every new individual is the product of the fusion of male and female reproductive cells, then each individual had to inherit two factors for the determination of a particular characteristic.

This was an important assumption. There were still many who believed that more than one sperm or pollen grain usually fertilized the egg.

Perhaps because he got tired of writing out "round factor," "wrinkled factor," or "tall factor," etc., or perhaps because of his mathematical background, Mendel developed a kind of shorthand as a means of expressing his ideas on paper. A dominant factor was represented by an upper-case letter such as "A." The corresponding recessive factor was represented by the same letter in lower case. For convenience, we will use letters which relate to English words for the characteristics. Using this system, the factor for round could be designated "R" and the factor for wrinkled as "r." A purebred round seed could then be written as RR; a purebred wrinkled seed as rr. Now, what of the round seeds that resulted from the round-wrinkled cross? Since each of these seeds received a factor from each parent they would be designated Rr. The hybrid Rr seed looks round because the R factor is dominant over the r factor.

With his system of notation Mendel could express his crosses on paper. This simple technique of letter notation enabled Mendel to visualize and understand what Darwin and a host of others had been unable to grasp. The cellular processes involved in the production of gametes were not to be discovered for another twenty-five years. Mendel actually described a basic principle involved in gamete production, when he expressed his crosses on paper with big and little letters. Mendel reasoned that since an individual has two factors for a characteristic, the factors must somehow separate when this individual formed gametes. Then, when the individual became a parent, it contributed one factor for each characteristic to its offspring.

Armed with his reasoning, he expressed the first cross of pure-breds as follows:

$$RR \times rr$$

The "X" means crossed with; RR designates a purebred, round seed plant and rr designates a purebred, wrinkled seed plant. Then he utilized mathematical probability to express the results of this cross. As stated before, Mendel theorized that the factors somehow separated. A gamete would then have one or the other factor. This supposition can be expressed as follows:

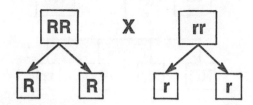

Now, what can happen? For convenience, we will call the RR on the left the male parent and rr on the right the female parent. In the process of fertilization, a sperm carrying an R can fertilize an egg carrying an r. The chances of this happening are 1 in 4, or one fourth. There is a one fourth chance that this R will combine with the other r. Following through, we see that there is again a one fourth chance that the other R in the male parent can combine with an r, and a one fourth chance that this R can combine with the other r in the female parent. If we add up all the one fourth probabilities we arrive at 1. Essentially, it all adds up to say that there is a 100 per cent chance that all the offspring of this cross will be hybrid Rr. As R is dominant all the seeds will look round. This is exactly the result obtained by Mendel.

Mendel obtained a 3:1 ratio of round seeds to wrinkled seeds in the second generation. There is no way of knowing what went through Mendel's mind as he wrote his letters and X's to express the results of his second cross. However, he could not have failed to have been excited by what he saw. The hybrid cross is expressed by:

$$Rr \times Rr$$

Again, for the sake of convenience we refer to the Rr on the left as the male and the Rr on the right as the female. Using mathematical probability again, we see that there is a one fourth chance of the male R combining with the female R, and a one fourth chance of the male R combining with the female r. There is a one fourth chance of the male r combining with the female R, and a one fourth chance of the male r combining with the female r. Summing up, we have:

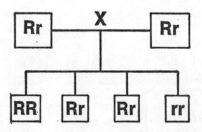

(Rr is the same as rR. Most geneticists always write the upper case letter first.) Remembering that R is dominant over r, this probability exercise shows us three fourths of the seeds should be round and one fourth should be wrinkled, or 3:1. As far as Mendel was concerned, the occurrence of the 3:1 ratio had been explained.

Mendel was now anxious to see what would happen if he observed two characteristics at a time. Would the 3:1 ratio still hold with a cross involving two contrasting characteristics? Does the presence of one factor affect the expression of another factor? These were some of the questions for which Mendel had to find answers.

For the next planting season, he selected seeds that were pure-bred for round, yellow seeds to be crossed with selected purebred, wrinkled green seeds. Mendel had determined that the yellow factor was dominant over the green factor. He planted the seeds, allowed the plants to mature, and, as before, carried out the laborious process of removing stamens and covering the flowers.

The seeds in the first generation were all round and yellow. In the next season, he planted these round, yellow seeds and allowed the plants to fertilize themselves.

Upon opening the pods and harvesting the peas, he found an assortment of seed shapes and colors. What he found was just the sort of thing that had confounded and discouraged earlier investigators. Mendel, however, was elated rather than discouraged with the results. He saw that there were four kinds of peas: round yellow, round green, wrinkled yellow, and wrinkled green. He harvested 556 peas. Three hundred fifteen were round yellow, 161 were wrinkled yellow, 108 were round green, and 32 were wrinkled green. The ratio was almost exactly 9:3:3:1. It was not at all difficult to see that the 3:1 ratio was still there. Taking each characteristic by itself, he saw that the ratio of round to wrinkled was 3:1 and the ratio of yellow to green was 3:1.

The results of the double hybrid cross revealed to Mendel that the inheritance of one set of characteristics had nothing to do with the inheritance of another set of characteristics. Each factor was inherited as a unit. Mendel had observed no blending of characteristics. He now felt that he had enough information to propose definite rules about inheritance. No one, including Mendel, knew it yet, but biology had become an exact science.

Mendel's rules of heredity can be summed up as follows:

1. Hereditary characteristics are determined by independent factors that act as units.
2. An organism inherits two factors—one from each parent—for each characteristic.
3. Some factors are dominant over others. When an organism inherits a dominant factor and a recessive factor, the dominant factor determines the characteristic.
4. The factors separate in the formation of gametes; they are unaffected by association with other factors.
5. The random distribution of factors in fertilization results in a predictable ratio of characteristics in the offspring.

Why did Mendel, the amateur, succeed where Darwin, the great scientist, failed? Darwin's lack of mathematical skills has been mentioned. It seems that Darwin could have overcome this, if not by himself, at least by consultation with someone more mathematically inclined. Many science historians think that Darwin was trying to "put the cart before the horse." He was too intent on finding the cause of variations before he knew anything of hereditary stability. How could he hope to arrive at any conclusions on how changes are transmitted in living things before he knew how stable characteristics are transmitted from one generation to the next? Darwin's approach to the problem may be compared to someone trying to fix a machine without knowing how the machine normally operates when it is not broken.

Mendel was sure of his observations. But perhaps because he knew that he was an amateur scientist, he felt that more testing was needed. Up to this point he had first carried out the crossing experiments and then worked out an interpretation with his symbols. He now proposed to predict the outcome first and then carry out the experiment.

Mendel still had some of the original purebred parent seeds. He had also saved some of the seeds of the first generation. From these seeds, he chose some showing the round-wrinkled and yellow-green characteristics. He knew that the parent round-wrinkled seeds were purebred. According to his notation, they can be written as RRYY (Y-yellow, y-green). The progeny of the RRYY × rryy cross he had established to be RrYy hybrids. Mendal now determined on paper the results of a RRYY × RrYy cross:

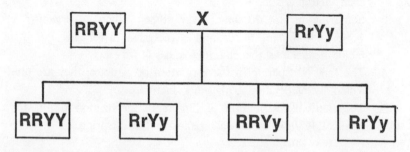

Mendel saw that in each of the four possible combinations there was a dominant factor. According to the rules he had proposed, all of the offspring of this cross had to be round yellow peas. Mendel planted the peas and carried out his crosses. When he opened the pods, he found what he had predicted—all round yellow peas.

The evidence for the validity of his principles was now stronger, but still he was not satisfied. Much depended on another cross he had carried out in a different part of the garden. This cross was essentially the reverse of the RRYY × RrYy cross. In this experiment, he crossed the purebred rryy (wrinkled green) peas with the hybrid RrYy peas. Mendel predicted the outcome as follows:

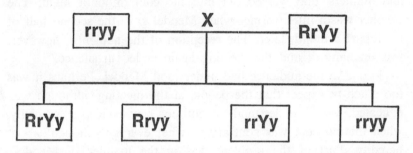

Mendel's notations indicated to him that this cross would result in four kinds of peas; round yellow, round green, wrinkled yellow, and wrinkled green in a ratio of 1:1:1:1. Because of an infestation of beetles, Mendel harvested only 110 peas. It was enough. He counted 31 round yellow, 26 round green, 27 wrinkled yellow and 26 wrinkled green. He had his 1:1:1:1. He was now ready to write his report. He devoted the last six months of 1864 to writing.

In the town of Brunn, where the monastery was located, some of the local intellectuals had formed an organization called the Brunn Society for the Study of Natural Science. As the name implies, this society devoted itself to studying and encouraging research in natural science. Mendel knew all the members of the society, and they all knew that he had been planting a lot of peas for the last six or seven years. In 1865 when the members of the society learned that Mendel had written a report of his work, they invited him to present his paper before a regular monthly meeting of the society.

It was very cold on the night the February meeting of the society was convened. There was a bit of bustle and stamping of feet as the doctors, engineers, chemists, astronomers, and other learned members of the society removed their coats and settled down to listen to Mendel deliver his report. Due to the length of the presentation, Mendel had been scheduled to give his report in two meetings a month apart. The members listened in silence as Mendel spoke of his factors, ratios, and principles. When he had finished, the chairman asked if the members had questions. There were no questions. The meeting was adjourned and the members of the society silently put on their coats, hats, gloves and mufflers, and walked out into the cold, moonlit night. The weather was a little warmer when Mendel gave the second half of his report a month later. The reception of the audience, however, was unchanged, and the meeting again ended in silence.

No one in the audience had understood Mendel. Perhaps it was too much to expect that the people at the meeting, although well educated, would understand an entirely new concept; especially, when this concept was contrary to what everyone thought about heredity. Part of the problem was in the manner of Mendel's presentation. He really knew he had discovered some of the basic principles of heredity, but he was half afraid to admit his own genius. He ended his presentation with a somewhat bland conclusion that color is expressed in the combination of independent characteristics.

Mendel was invited to publish his paper in the society's small journal. The paper, entitled "Experiments in Plant Hybridization," was published in 1866. About 120 universities and scientific organizations regularly received the Brunn journal. Mendel, of course, hoped to receive some communications about his paper. He especially hoped to hear from some botanists. He waited and waited, but he received no mail relating to his work. Mendel must have felt very much alone.

He hungered to talk about his experiments with someone who might appreciate their significance. There was certainly no one in Brunn. He could not understand why there had been no comment, positive or negative, from anyone who might have read his paper. He thought that if his work was read and upheld by someone who was an established expert in botany, it might then be brought to the attention of people interested in heredity.

He then decided to write to one of the most respected botanists of the time, Carl Naegli. On New Year's Eve of 1866, Mendel composed a long letter and enclosed a copy of his paper. In the letter, Mendel used such terms as "your honor" and "your eminence" in referring to Naegli. Mendel waited many months before he received an answer. The letter was somewhat disappointing. Apparently Naegli, too, had missed the point. He complimented Mendel on the exactness of his work, but went on to say that much more work had to be done before the nature of inheritance was known. Mendel wrote to Naegli again. His correspondence with the great man extended over seven years.

Naegli's letters were patronizing. They read like a teacher's advice to a boy who was preparing a project for his first science fair. Mendel was not insulted; he was grateful to hear from the great man. He even took Naegli's advice to stop experimenting with peas, and to do some work with a plant called hawkweed. Unknown to both Mendel and Naegli, hawkweed was unsuitable for breeding experiments. The results that Mendel obtained with hawkweed were so incomprehensible, that they made Mendel doubt the validity of his principles. The problem was that hawkweed seeds could develop from unfertilized female gametes. Since Mendel's principles were based on inheriting factors from two parents, the hawkweed results were disastrous to Mendel's faith in his theories.

Why did Mendel not write to Darwin? Mendel certainly knew of Darwin's theory. Mendel's well-thumbed and marked copy of *Origin of Species* still exists in the Mendel Museum at the Brunn Monastery. It is even more difficult to believe that a man of Naegli's stature could carry on a correspondence with Mendel for eight years and fail to understand the significance of Mendel's work. There are some who believe that Naegli used Mendel's work and claimed it as his own. There is little evidence to uphold this view. Naegli continued to propose his own (later proved wrong) ideas of inheritance. These ideas were formulated long before he knew Mendel.

In 1868, Mendel was elected Abbot (head) of the monastery. Mendel's new responsibilities kept him from his experiments. He tried to continue, but as the years went by he had less and less time for plant breeding. He stopped writing to Naegli in 1874.

Mendel became very active in community affairs, even serving as chairman of a bank. But he was still interested in science, did some experiments in meteorology, and found time to raise some prize fruit trees and vegetables. He raised bees and even replanted an entire bare mountainside to provide pollen and nectar for his bees. He apparently never again spoke of his pea experiments to anyone, even though he attended conventions of plant breeders in many European cities.

When he died in 1884, the entire city of Brunn mourned. He had become the most revered and respected person in the city. Very few people outside of Brunn even knew he had died. Indeed, in the year he died, most people in the world never knew he had lived.

A new century was to begin before the world was to know of the work done by the modest monk. Mendel did his work in the nineteenth century, but his legacy belongs to the twentieth century. Darwin and other searchers of the nineteenth century labored under the restraints imposed by the past, never knowing that the future was already there, gathering dust on a library shelf.

Mendel never knew just what his "merkmalen" were. He had an idea that they must be in cells. During Mendel's lifetime, men looking into microscopes were finding out a lot about cells. They didn't find the merkmalen either, but they pointed out where to look.

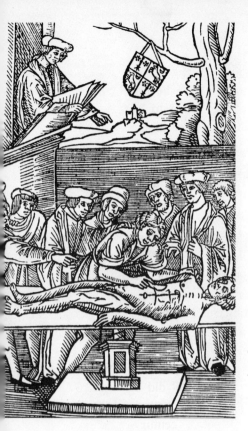

1. The professor reads from Galen while the barber cuts as directed by the professor's assistant

A figure from Vesalius' Fabricia *De Corporis Humani Fabricia*

3. Carl Linneaus

4. The Comte de Buffon

5. An illustration from *Historie Naturelle* depicting a scene from a laboratory of the Enlightenment period

6.   The Baron Cuvier

7. Erasmus Darwin   8. Jean Baptiste de Lamarck

9. H.M.S. *Beagle* in the Straits of Magellan

10.   "A most singular group of finches"

11.   Darwin's study at Down House

ON

THE ORIGIN OF SPECIES

BY MEANS OF NATURAL SELECTION,

OR THE

PRESERVATION OF FAVOURED RACES IN THE STRUGGLE
FOR LIFE.

BY CHARLES DARWIN, M.A.,

FELLOW OF THE ROYAL, GEOLOGICAL, LINNÆAN, ETC., SOCIETIES;
AUTHOR OF 'JOURNAL OF RESEARCHES DURING H. M. S. BEAGLE'S VOYAGE
ROUND THE WORLD.'

LONDON:
JOHN MURRAY, ALBEMARLE STREET.
1859.

The right of Translation is reserved.

12.   The title page from *Origin of
Species*

13.   Charles Darwin and Thomas Huxley

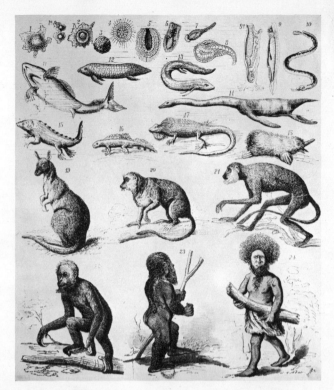

14. One of Ernst Haeckel's evolutionary charts—many of the pictured animals are completely fictitious

Contemporary cartoon by Thomas Nast.

"Gorilla: 'That Man wants to claim my Pedigree. He says he is one of my Descendants.'"

"Mr. Bergh [founder of the A. S. P. C. A.]: 'Now, Mr. Darwin, how could you insult him so?'"

15. This cartoon, drawn by the American cartoonist Thomas Nast, shows how widely Darwin's ideas were misunderstood

16.   Gregor Mendel

17.   Mendel's garden. A statue of Mendel is to the left of the garden

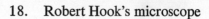
18.   Robert Hook's microscope

19.   Nachet's "multiple microscope" manufactured in 1856

20.   Rudolph Virchow

21.   Auguste Weismann

22.   This cartoon, published in 1883 is an indication of the great interest in heredity before the "rediscovery" of Mendel's work. The caption reads, "The great choice of ancestors. What atavistic influence goes to determine this?"

4. THROUGH THE LOOKING GLASS

Darwin knew little, if anything, about microscopes and what could be seen with them. He was too busy trying to find answers to his own problems. He owned microscopes and may have used them from time to time. But he did not think that the microscopic structure of cells and tissues had any bearing on his species problem.

In the 1840's and 1850's when Darwin was sorting out evidence for natural selection, a growing number of biologists were turning to microscopic studies. They all knew about Darwin and his work, certainly, but it does not seem that many of them thought their work was related to problems of species.

The microscopists of the time were slowly becoming convinced that all living things were made of cells and that cellular structure was an important principle of the unity of life. Of course, there were many who did not think that cells were particularly significant. Darwin remained aloof from the microscopists' controversy with such statements as, "I have not attended much to histology." Histology is the name given to the microscopic study of cells and the tissues they compose.

Many of Darwin's colleagues were more directly involved. Darwin's defender, T. H. Huxley, wrote in 1853 that cells ". . . are . . . no more the producers of vital (life) phenomena than the shells . . . along the sea-beach. . . . Like these, the cells mark only where the vital tides have been and not how they have acted." Darwin, who had faced more than his share of controversy, was very careful to avoid the cell question. When he proposed his pangenesis hypothesis he used the word cell sparingly, even though he thought the gemmules were produced in the cells.

Cells were not exactly new to biologists in the mid-nineteenth century. The word "cell" in its biological sense had been coined in 1665 by Robert Hooke, the curator of the Royal Society. He had used the word in his description of a thinly sliced piece of cork as seen through an interesting new instrument called the microscope. The Royal Society had purchased a number of microscopes for use by its members. While most of the gentlemen merely amused themselves with this fascinating toy, Hooke used it for serious study. In 1665, he published a book called *Micrographia*, which was an account of his observations, complete with many detailed drawings.

Hooke observed many familiar substances with his microscope, which had a magnifying power of about thirty diameters. In his description of cork, he explained that he had cut a slice of cork "exceeding thin" and observed that the cork seemed to be made of "pores or cells [which] were not very deep but consisted of a great many little Boxes." Hooke attached no more significance to this observation than that the structure explained the sponginess and buoyancy of cork. Almost 200 years were to pass before biologists began to understand the significance of the "many little boxes."

There were few people in Hooke's time and in the immediate following years who thought the microscope would ever be an important tool in the study of biology. This attitude is understandable. Although looking through a seventeenth- or eighteenth-century microscope was a fascinating experience, the quality of the image was more psychedelic than informative. The images were distorted and ringed with halos of color. A person with imagination could "see" whatever he wanted to see. It is no coincidence that the growing interest in cellular structure from about the 1830's on came with the development of better microscopes.

The problems of early microscopes were the problems of the physical behavior of light when it passes through a lens. A microscope is basically a lens or a system of lenses. Lenses usually are made of glass, but they can be made of almost any transparent material. A drop of water can be a lens. Indeed, the first lenses were water. The Roman philosopher Seneca wrote in A.D. 65 that glass globules filled with water were useful in seeing things that "frequently escape the eye."

When light passes through a substance, that substance is called the medium. When light passes from one medium to another something happens to it. Air is the usual medium through which light passes before it gets in our eyes. Glass is a thicker medium than air. When light passes from air through the glass of a lens, the rays of light are bent. The image produced by the bent light can be magnified, reduced, or distorted. The degree of magnification, etc. is dependent on many things including the shape of the lens, what the lens is made of, and its distance from the eye and the observed object.

Distortion of the image was a serious problem with early microscopes. There was another problem equally as bothersome. When white light passes through something like a lens, the light is broken up into the colors of the spectrum. Early microscopists saw brilliant displays of colors when they looked through their instruments. Beautiful though this may have been, this phenomenon, called chromatic aberration, interfered with meaningful observation.

As far as is known, the first microscope was made in 1590 by two brothers, Jan and Zacharias Janssen, in Holland. Their microscope consisted of two lenses placed at opposite ends of a lead tube. Since the Janssens' microscope had more than one lens, it is called a compound microscope. The microscope Robert Hooke used was a compound microscope.

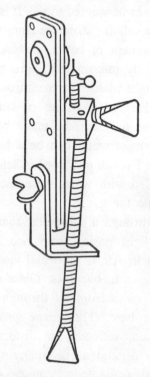

LEEUWENHOEK'S MICROSCOPE

Some success with single-lensed, simple microscopes was achieved by a cloth dealer in the Dutch town of Delft. This cloth dealer, Anton van Leeuwenhoek, had become rather skillful in the art of grinding lenses. By the 1680's he had constructed several microscopes. His microscope was a single lens mounted in a brass plate. The object of study was placed on a needlelike affair mounted on one side of the lens.

Leeuwenhoek achieved fairly high magnifications with his microscope. He kept detailed records of his observations so it is known that he saw bacteria, although he did not know what they were. He also saw human sperm cells and single-celled animals. He called the one-celled animals "cavorting beasties." He had no idea that the cavorting beasties and the bacteria were one-celled organisms.

Leeuwenhoek communicated with the Royal Society over a period of fifty years. He was eager to tell everybody what he saw, but he would say nothing of how he managed to see what he did with his simple microscope. Part of the answer was that his lenses were better than anyone else's. At the time, most lenses were made by melting glass. The results were inconsistent. Leeuwenhoek patiently ground his tiny lenses. On one occasion, he ground a lens from a grain of sand.

Many of the early microscopists used compound microscopes and the difficulties of using them are evident in some of the fantastic accounts of "observations" made with these instruments. One such series of imagined observations was the homunculus, or little man, in sperm cells. The controversy between the spermists and the ovists followed.

The arguments which raged between the spermists and the ovists contributed little to the development of biological science. But they were part of a much more significant controversy which was not to be completely resolved until almost the end of the nineteenth century. This controversy centered about the question of whether organisms developed through a series of changes into mature forms, or whether they merely expanded from a preformed miniature version. Both the spermists and the ovists proposed the latter view. Workers interested in this problem developed a branch of biology called embryology.

Throughout most of the eighteenth century, there were few, if any, biologists who thought Leeuwenhoek's cavorting beasties and Hooke's little boxes had anything to do with each other. Most of the early microscopists were interested in either embryology or the structure of plants. Microscopists had to work with poor microscopes, and there was a short supply even of these. There were few business people who thought that the manufacture and sale of microscopes was likely to be a profitable enterprise.

Nehemiah Grew, another curator of the Royal Society, devoted his entire life to the study of plant structure. Both he and Hooke observed little boxes in plant tissues other than cork. Neither Hooke

nor Grew, however, suggested that this structure was common
to all plants. Grew was imaginative in the descriptive terms he
used. He may have been tired and thirsty when he wrote that the
internal structure of certain plant stems reminded him of "the
froth of beer."

What to call things was a serious problem among biologists.
In the scientific papers of the seventeenth and eighteenth centuries,
there are references to observations which may or may not have
been descriptions of cells. Terms such as "globules," "bladders,"
and "pores" probably referred to what are called cells today.
Other investigators used the word cell to describe things that were
not cells. In many instances what were described as cells may
have been stray circles of light or air bubbles.

A German embryologist named C. F. Wolff seems to have come
closer than anyone else to proposing a general cell theory in the
eighteenth century. In 1759, he wrote:

> . . . the particles which compose all animal organs in their earliest
> inception are little globules which may be distinguished under a mi-
> croscope.

Wolff had made extensive studies of the embryology of the chick.
He was violently opposed to the preformationists. He had seen
how groups of globules that looked nothing at all like a chick
developed into something which very definitely was a chick.

Of course, the preformationists did not believe him. They had
no reason to do so, and they were just as convinced of their
own observations. Reporting observations was another big problem of early microscopists. Every observer had to be taken at his
word. There were no cameras and everything had to be drawn.
Not all biologists had equal artistic talents. In the splashes of
color and distortions of an eighteenth-century microscope one
could "see" almost anything. If the observer was looking for
something to uphold his own particular hypothesis or idea, it was
not hard to convince himself that he saw whatever he needed to
see. A little artistic license in drawing probably helped to "uphold" many a microscopist's claimed observations.

It is difficult to say whether the improved microscopes of the nineteenth century were the direct cause of the intense interest in the study of cells at the time, or whether the intense interest in cytology was the cause of the development of better microscopes. The answer to the troublesome chromatic aberration had been determined as early as 1735 by Chester Moor Hall, a telescope maker. He combined lenses of different shapes and kinds of glass in such a way so that the various lenses "canceled out" the colors. As in the case of many innovators, no one paid too much attention to him. In 1760, another telescope maker named James Dolland applied the same principles with more successs. He managed to sell some of his telescopes. The colors were not entirely eliminated, but the improvement over earlier telescopes was obvious. This "achromatic" (without color) telescope was soon in great demand.

It was immediately known that the same principles could be applied to microscopes, but there were many technical problems. Microscope lenses are much smaller than telescope lenses. In order to achieve high magnifications they must be steeply curved, but it was very difficult to fit these small, curved lenses together. An achromatic microscope was made in 1791 by Francois Beelsnijder of Amsterdam, Holland, a cavalry officer. It is not clear just why Beelsnijder took the trouble to make this microscope, but a cavalry officer may have had just as much reason to make a microscope as a cloth dealer did.

By the early nineteenth century, there were a few manufacturers in France and Germany offering achromatic microscopes for sale. This coincided with a tremendous increase in the amount of work being done in embryology and the study of animal structure.

The first steps toward solving another problem of microscopes began around 1830. This was the problem of distortion called "spherical aberration." Again the solution lay in the arrangement of the lenses. Manufacturers strived to outdo their competition in meeting the demand for better microscopes. By the 1840's there was a good supply of fairly high-powered instruments which had much less aberration than earlier low-powered microscopes of single lens and compound design. Some firms in France made achromatic microscopes so cheaply that they were bought for use in schools by entire classes of students.

The manufacturers made another significant improvement. They offered microscopes of much higher resolving power than those of the eighteenth century. The resolving power of a microscope is the minimum distance between two points which can be seen separately through the microscope. Lack of resolving power limited the usefulness of earlier instruments that had high magnifications. As magnification is increased, resolving power tends to go down. This is still a problem today with low-quality microscopes and telescopes. The development of microscopes with high resolution made it possible for investigators to see more of the internal structures of cells.

The early nineteenth century also saw advances in the techniques for preparing material for microscopic study. In order for anything to be studied with most microscopes, it must be thin enough to be transparent. Recall that Hooke had sliced his cork "exceeding thin." Most microscopists of the eighteenth century had prepared materials by squashing or smearing them on pieces of glass. Cells would frequently come apart under such treatment. In the late eighteenth and early nineteenth centuries many investigators made devices that could cut tissue into very thin slices, much thinner than Hooke would have ever dreamed. Later, improved versions of these devices were called "microtomes." Others found that treating the tissue with certain chemicals and coloring them with various staining materials would make observations easier. These techniques enabled investigators to see parts of cells which do not stand out from the general structure of a cell in its natural state.

By the 1830's, many investigators were convinced that something like cellular structure was a general characteristic of plant tissues. Not as many thought the same of animal tissues. This was probably due to the fact that many animal tissues are less obvious in cellular structure than plant tissues. The cell limits are not as distinct.

There were two general schools of thought about the cells, globules, etc. of plant tissue. One group believed that the cells were just spaces between fibers which make up the bulk of plants.

The second group felt that the cells were separate entities which make up the tissues of the plant. An important piece of evidence was obtained for the latter school by Trevianus in 1805. He actually succeeded in separating the individual cells of plant tissue.

Lamarck was an early believer in the fact that animals have a cellular structure. He felt there was evidence for him to propose that much of the structure of animals "are in general the productions of cellular tissue." This pronouncement, in 1809, caused no particular excitement. Not too many people listened to what Lamarck had to say about anything.

A strong case for cellular structure of plants was provided by a French physician, J. Dutrochet. Dutrochet was an ardent "globulist." In 1824 he separated the individual cells of a mimosa plant by boiling some of the plant tissue in nitric acid. In his report of this experiment, he discussed the difficulties of using the microscope and proposed that new methods of investigation were needed. In the same year he was sure enough of his evidence to propose that all tissues, animal and plant, are actually "globular cells of exceeding smallness" and that animal and plant organs are "actually only a cellular tissue variously modified."

Improved microscopes and techniques encouraged investigators to look at the internal structure of cells. One of the best observers of the period was an Englishman, R. Brown, another physician turned botanist. In 1823 he published, in the *Transactions of the Linnaean Society,* an account of his observations of the microscopic structure of orchid plants. He reported observing in each cell ". . . a single circular areola, generally somewhat more opake [sic] than the membrane of the cell. . . ." He was as meticulous in his writing as he was in his observation. He stated that he first observed the "areola" in epidermal or surface tissue of the plant. He went on to say that "This areola, or nucleus of the cell as perhaps it might be termed, is not confined to epidermis, being found not only in the pubescence of the surface . . . but in many cases in the parenchyma or internal cells of the tissue. . . ." He then listed other plant tissues in which he had observed what he termed the "nucleus of the cell." He even cautioned the readers not to mistake a structure observed in pollen grains as the nucleus. He suggested that this structure, likely to be mistaken for a nucleus, was the beginning of the pollen tube, a feature of pollen which had been observed some ten years earlier.

By the 1830's, the evidence that all living things were made of cells was overwhelming. An accumulation of knowledge which spanned almost 200 years had convinced many disbelievers. There were still many who did not think that cells were common to all living things. An idea or hypothesis does not become a generally accepted theory as a result of someone's dramatic announcement. But if someone who has earned a reputation accepts it, general acceptance of the idea is helped along. The theory may, in time, be replaced by a theory which explains observed phenomena in a better way, or it may gain complete acceptance and withstand all attacks against it. It may then be referred to as a law or axiom.

The name of Matthias Schleiden has been frequently associated with the establishment of the cell theory. Matthias Schleiden was a well-known natural scientist in the 1830's. He was unusual among natural scientists in that he had been a lawyer, but he did not do very well in the legal profession. He was so despondent at his failure in law that he attempted suicide. After his recovery, he devoted his attention to natural science. In 1842, he published a book, *Principles of Scientific Botany,* in which he vigorously attacked a group of botanists in Germany who had apparently fallen back on the methods of Aristotle in the way they studied botany. This group, the "Nature Philosophers," felt that there was nothing to be learned from investigating the parts of plants. They believed more was to be learned from studying the whole plant, and that tearing a plant apart for study was "unnatural."

In his book, Schleiden maintained that botanists had to study the structure of cells and the embryonic development of organisms. Schleiden defended his positions strongly, but he was not always right, and he was reluctant to change his mind about anything no matter how much evidence was presented. Schleiden did not "discover" the cell theory, but statements he made in 1838 were very important to the acceptance and extension of the theory.

Another important contribution to the acceptance of the cell theory was made by Theodore Schwann. In 1839, he published a book in which the words "cell theory" appeared for the first time. Schwann had done extensive work on animal cells. He not only applied the cell concept to animals but saw important similarities between animal and plant cells. Schwann suggested that cells were somewhat independent in their activities. But he felt they were controlled by the whole organism.

Schwann thought that the observation of nuclei in the cells of both plants and animals was extremely important. In Schwann's biography there is the following account of a conversation he had with Schleiden:

> One day, when I was dining with M. Schleiden, this illustrious botanist pointed out to me the important role the nucleus plays in the development of plant cells. I at once recalled having seen a similar organ in the cells of the notochord [a structure made of cartilage found in animals of the phylum Chordata], and in the same instant I grasped the extreme importance that my discovery would have if I succeeded in showing that this nucleus plays the same role in the cells of the notochord as does the nucleus of plants in the development of cells.

Schwann believed that the cell was the functional as well as the structural unit of all living things. His excitement at finding that nuclei were in the cells of both plants and animals is understandable. He had found an indication of a unifying principle of life.

Schleiden's remark that the nucleus was important in the development of cells related to the next line of research. There were many natural scientists who were not yet convinced of the cell theory, but those who were convinced directed their attention to the question, "Where do cells come from?" A controversy immediately developed; however, almost everyone thought that the nucleus had something to do with the production of new cells in an organism.

Dutrochet had stated that the growth of an organism was due to an increase in the number of cells. Many others had the same idea. There was general agreement that as an organism increased in size, the number of its cells increased. The arguments were over just how new cells came into being. Schleiden believed that the nucleus was a sort of immature cell which blossomed out into a full-sized cell. Other workers thought new cells came from broken pieces of old cells. Some people thought new cells grew in the spaces between cells and pushed the older cells out of the way. Still others thought new cells came from old cells splitting into two equally sized new cells. Many believed that cells sort of "hardened" from a general liquid produced by the organisms. The controversy quickly boiled down to Schleiden's views versus the equal division concept.

Schleiden had two views of the origin of cells. In addition to his ideas on the expansion of the nucleus, he thought that some cells formed from a general noncellular liquid that was formed first. Schleiden based his hypothesis on observations of tissue within the ovaries of plants. This part of seed plants, called the embryo sac, is where the egg cells of the plant are fertilized and develop into seeds. The embryo sac was an unfortunate choice. It happened to be one of the few instances where cells are formed without cell walls. The walls form later. To an observer, it does look as though cells are forming from a liquid mass. Schleiden elaborated his hypothesis by saying that granules formed in the tissue which expanded into nuclei and then into cells. He did not extend his observations to match his expanded hypothesis.

Schleiden's forceful personality and a large group of devoted students did more to spread Schleiden's ideas than any evidence he presented. Schwann went along with Schleiden although he had some doubts. He tried to apply Schleiden's ideas to the formation of animal cells, but couldn't.

Schleiden was widely respected, but his view could not stand up to the evidence for equal cell division. Carl Naegli went along with Schleiden only in respect to the embryo sac, but in all other instances, Naegli said that cells arose from equal cell division.

At about the same time, Schleiden and Schwann championed the cell theory, a group of workers had been investigating the contents of cells. It was found that the semiliquid material in cells was basically the same in both plant and animals. This material was most easily seen in organisms such as amoebae.

The stuff of which the amoeba is made was called "sarcode" by a Frenchman, Felix DuJardin, in 1835.

A Czech physiologist, Purkinje, used the word *protoplasm* to describe the cell contents. It was apparent that Hooke's cells had not been living cells but the walls of dead cells.

While the sarcode, protoplasm, etc. was most easily seen in organisms such as amoebae and paramecia, it was, at first, not clear whether these *Protozoa*, as they were called, were one-celled or many-celled. In 1845, a German worker, von Siebold, conclusively demonstrated that protozoa were single-celled animals. The cavorting beasties of Leeuwenhoek and Hooke's little boxes had met on common ground.

The name of Rudolph Virchow is generally associated with the theory of new cells arising only from preexisting cells. Actually, he did not provide a great deal of experimental evidence for this view, but he was active in advertising and spreading the doctrine. As so many other biologists, he started out as a student of medicine. While in medical school he became interested in pathology, the study of diseased tissues, and made this study his lifework. His graduate work was done in Berlin in 1848 when revolutions were breaking out all over Europe. He was active in revolutionary causes in Berlin, and found it necessary to leave that city in order to avoid being jailed.

While in Berlin, he started a journal which is still published under the name of *Virchow's Archiv*. He agreed with Schleiden's view on cell formation, and many articles upholding Schleiden were published in the *Archiv*. After his forced exit from Berlin in 1848, he taught pathology at the University in Wurzberg, until the political situation eased, allowing him to return to Berlin in 1856.

During his stay at Wurzberg, he did a great deal of work with the microscope. The accumulated observations during this period made it impossible for him to continue to agree with Schleiden, and in 1855 he published a paper on the subject of cell formation. In this paper, he summed up his views in a Latin phrase which has come to be one of the most frequently quoted statements in biological science, *omnis cellula e cellula* ("all cells come from preexisting cells"). Virchow had nothing to offer on just how cells divide equally; that was twenty-five years away.

Biologists did not come flocking to pay homage to Virchow just because he was adept at coining Latin phrases. The controversy continued, more bitter than it had ever been. Schleiden said some very unkind things about his ex-friend Virchow. Perhaps Schleiden was embittered because many of his own students had adopted the equal division hypothesis. Virchow continued to urge medical doctors to make more use of the microscope. He proposed the idea that cancer and the lesions of tuberculosis were the results of the abnormal division of cells.

About a year after Virchow published his famous Latin phrase, Gregor Mendel started to plant peas. There is nothing to indicate that Mendel was prompted to start his work by Virchow's writings. There was little in the writings of anyone at the time which could have encouraged the completion of work such as Mendel's. That he was able to begin anything at all is a tribute to his genius.

When Mendel completed his experiments, his conclusions rested upon the acceptance of certain basic assumptions. One of these was that plants are sexually reproducing organisms. The second was that the pollen grain was the male gamete and that only one male gamete fertilizes one egg cell or ovum. (Actually, the male gamete as such is within the pollen grain, but this fine point of detail did not affect Mendel's conclusions.) The first of these assumptions had been accepted since the time of Linnaeus, but the others were by no means fully accepted when Darwin and Mendel did their work.

The state of knowledge about fertilization at that time can best be described as confused. There was a vague idea that sperms had something to do with stimulating animal ova or egg cells to start development. Even less was known about the sexual reproduction of plants. Botanists and zoologists, armed with improved microscopes, turned their attention to these problems. In the process, they managed to point the way to finding out more about just how cells divided to produce new cells. What they saw in reproducing cells was to show another generation of biologists just how fantastic a genius Mendel had been.

Much of what little that was known about plant reproduction had been provided by an Italian botanist who had been fascinated by some bulging grains of pollen. Jean Baptista Amici (1784–1860) couldn't seem to make up his mind about what he wanted to do. When he engaged in microscopical work, he saw the necessity for improving microscopes. He would then spend years working on improving microscopes. When improvements were made, Amici would then be anxious to return to microscopical research with the better instrument. In the course of this seesaw career, he managed to contribute a great deal both to the development of better microscopes and the knowledge of plant reproduction.

In 1823 while observing some pollen grains, he observed that a tubelike projection was extending from one of the grains. Three years later he saw the same thing with pollen grains which had been placed on the pistil of a flower. He observed that the tube from the pollen grain extended all the way down to the lower part of the pistil, called the ovary.

At first, neither Amici nor others who had observed this phenomenon knew what it was all about. Schleiden offered his views. He said that the pollen tube was the female element of the plant. Amici did not agree with his view. He expressed the idea that the female element of the plant was in the ovary before the entrance of the pollen tube. Schleiden did not like to be contradicted and was very abusive to Amici.

Amici continued to make observations, and by 1846, he was able to show that an egg cell or ovum developed in the plant ovary and was stimulated to grow into a seed by the presence of the pollen tube. Others confirmed Amici's observations, but no one knew just why a tube from a grain of pollen should stimulate the growth of the plant ovum into a seed.

The animal embryologists were not idle during this period. In 1844 two French workers, Prevost and Dumas, overthrew an idea which had caused great confusion for more than forty years. The source of the confusion was an experiment carried out in 1780 by Lazzaro Spallanzani, an Italian biologist of outstanding ability. Spallanzani wanted to find out if spermatozoa were necessary for the fertilization of frogs' eggs. He passed the seminal fluid of the male frog through a fine filter, hoping to filter out the spermatozoa. He then exposed frogs' eggs to the filtered seminal fluid. The eggs thus exposed started to develop into tadpoles. From these observations, Spallanzani concluded that spermatozoa were not necessary for fertilization.

Prevost and Dumas, in repeating the experiment, varied the procedure. They succeeded in separating the spermatozoa from the seminal fluid. Then they exposed the frogs' eggs to the spermatozoa without the seminal fluid. The eggs were fertilized and developed. Apparently Spallanzani's problem was that the filter he used was not fine enough.

Prevost and Dumas saw the spermatozoa penetrate the jellylike coating of the frogs' eggs. But they did not see the spermatozoa actually enter the eggs. In the ten years following the work of Prevost and Dumas, it was shown that spermatozoa were cells and were produced by other cells in the body of the male animal.

There was a general idea that pollen grains and sperm cells were both male gametes but there were few ideas as to just how the sperm or pollen grains, stimulated the start of embryonic development. It was not clear if only one, or several, male gametes were necessary to fertilize the female gamete. In 1868 Darwin stated that several sperm were necessary for fertilization. Very little, if anything, was known about the fertilization of eggs in mammals. In 1843 when a certain Martin Barry reported that he had seen fertilized ova in the reproductive organs of a rabbit, no one believed him.

Mammals, birds, and other animals whose egg cells are fertilized internally were not usually sound sources of research material. Invertebrate animals proved to be good subjects for study because the eggs are produced in massive numbers and are normally expelled from the animal's body in an early stage of development. A parasitic worm, called the Ascaris, proved to be a valuable organism for biologists. This species, a parasite of horses, was easily obtainable from stables, where they were removed by the thousands from horse manure. The eggs were relatively large, easy to observe with the microscope, and were produced by the tens of thousands by every female Ascaris.

In the 1870's biologists working with the eggs of Ascaris observed that in a just-fertilized Ascaris egg there were two nuclei, while there was only one nucleus in an unfertilized egg. The two nuclei in the fertilized egg were seen to fuse just before the eggs started to divide into the many cells which eventually developed into the larval worm. The source of the second nucleus and the significance of the fusing was a mystery.

Oscar Hertwig worked with an improbable creature called the sea urchin. A relative of the starfish, it looks like a flattened ball covered with actively moving spines. A picture of a sea urchin, enlarged and projected on a screen, would make a respectable monster for a horror movie. Hertwig, however, regarded it as no monster, but as an excellent source of eggs and sperm. Male and female sea urchins shed their gametes in sea water where they are fertilized. They produce gametes by the millions.

The entrance of the sperm cell into the egg cells of the sea urchin was easy to observe. Hertwig saw that when a sperm cell entered an egg cell, the sperm cell head became a nucleus. The mystery of the two nuclei had been solved, at least as far as sea urchins were concerned.

A colleague of Hertwig, Hermann Fol, worked with starfish. In 1879 he observed that one, and only one, sperm fertilized each egg. This was fourteen years after Mendel presented his paper to the Brunn Society. If Fol or someone else had demonstrated the one sperm-one egg idea before, perhaps Mendel's paper would have been understood in Mendel's lifetime.

The 1880's and 1890's saw an explosion of cell research. Never before had so many men been so actively engaged in one aspect of biological science. These men began to call themselves cytologists. The journals were flooded with papers. Workers had to wait months for their papers to be published. One thing that contributed to this situation was an important improvement in microscopes, announced in 1877. This was the oil immersion objective (lens), invented by the German microscope maker, Ernst Abbe. This system of lenses effected a very high magnification. In order to be used, a drop of clear oil or other clear substance is placed on top of the slide. The tip of the objective is lowered into the drop. The physics of this is very complicated, but basically what it did was to eliminate one of the changes in refractive index of the substances through which the light passes before it reaches the eye of the observer. The drop of oil has about the same refractive index as the lens. More light could enter the objective at a better angle to provide excellent resolution. Abbe also invented a *condenser* which concentrated the light entering the objective.

The growth of interest in the study of the cell was not due to the oil immersion objective alone. Most of the impetus came from the knowledge that the study of the cell was extremely significant to all branches of biology. The cell had been determined to be the structural and functional basis of all living things. As such, greater understanding of the cell would lead to a greater under-standing of all life's phenomena.

Most of the cytological work was done in Germany. New in-formation in embryology, the contents of cells, and the reproduction of cells was rapidly gathered. So much work was being done that it was almost impossible for one man to keep up with all the new developments.

The idea that cells split into two generally equal new cells was well accepted by the 1880's. Embryologists observing the fusion of the sperm and egg nuclei had noted that the fertilized egg started to divide almost immediately following the fusion of the nuclei. The oil immersion objective and staining techniques enabled the worker to see many fascinating but puzzling things in dividing cells. They reported seeing star-shaped structures which radiated fibers in all directions in dividing cells. The nucleus was observed to disappear during cell division only to mysteriously reappear in the new cells, usually called daughter cells.

Not all the work was done in Germany. A very active French worker, E. G. Balbiani, contributed a great deal. He studied protozoa for a good many years, but for a long time did not understand that they were one-celled organisms. In 1876 Balbiani was working with cells from the ovaries of grasshoppers. He reported that he saw what he called *batonnets étroits* (little narrow sticks) form from the nucleus of the dividing cell. He went on to say that the *batonnets* divided across in the middle and became a mass which turned into a daughter nucleus as it moved into each daughter cell.

Other workers observed the *batonnets,* but referred to them by a variety of names. They were observed only during cell division and assumed a dark black color upon the addition of certain stains.

No one person can be credited with determining all the details of cell division. However, if one had to be chosen, Walter Flemming would be a strong candidate. Flemming worked with cartilage cells from the embryos of salamanders. These cells are actively dividing and proved to be very good material. He also saw the *batonnets,* but he called them "threads." In 1879 Flemming published a detailed description of the division of salamander cells. He proposed nine definite phases in the division process. He described the nucleus that seemed to fade away and the appearance of coiled threads where the nucleus had been. These threads continued to coil and become thicker. He commented that the threads only could be seen with proper staining. He proposed that the coiled, thickened threads split lengthwise rather than across as Balbiani had proposed. Flemming went on to describe how the split threads moved away from each other. He described the actual division of the cell by "pinching" and the formation of new nuclei from the groups of threads which moved away from each other into each of the two new cells.

Similar events of cell division were observed in the cells of other animals and plants. The formation of the threads and their movements were seen in many plants and animals. A name was needed for so general a phenomenon. A worker named Schleicher proposed "karyokinesis" (nuclear movement). Flemming proposed a derivative of the Greek word for thread, *mitos.* Adding "is" to the Greek word, he derived the word, *mitosis.*

Eduard Strasburger also had made many observations of dividing cells. In his observations, he proposed names for the various phases, such as *prophase* for the early phases, *metaphase* for the middle phases, and *telophase* for the final phase. He called the

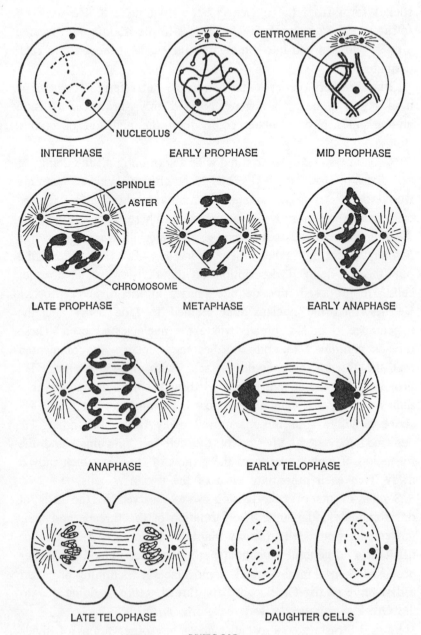

INTERPHASE EARLY PROPHASE MID PROPHASE

NUCLEOLUS CENTROMERE

SPINDLE ASTER CHROMOSOME

LATE PROPHASE METAPHASE EARLY ANAPHASE

ANAPHASE EARLY TELOPHASE

LATE TELOPHASE DAUGHTER CELLS

MITOSIS

batonnets, threads, etc. *chromatin,* from the Greek word for color. A few years later, a worker named Waldeyer used the term "chromosome" (body of chromatin). This term was generally accepted.

In the course of investigating mitosis, many more intriguing things were observed about chromosomes. Many observers thought that in any given species the number of chromosomes seen in dividing cells was always the same. Many proposed that chromosomes had particular sizes and shapes, usually recognizable. This was difficult to prove in most organisms because of the large number and small size of chromosomes. A question arose as to whether the chromosomes broke up between divisions or stayed together and assumed a stringier, more drawn-out form. This question was not to be resolved until the next century.

Again, the Ascaris proved to be a very useful organism for the cytologist. It was found that its cells had very few, relatively large chromosomes. One variety was found with four chromosomes and another was found with only two chromosomes. The consistency of the size and shapes of the chromosomes from one division to another was very easy to see in the cells of the Ascaris. Of course, this observation did not necessarily apply to other organisms.

The clearly seen chromosomes of the Ascaris showed something else that had been suspected but never clearly observed in other material. The gametes of the Ascaris had one half as many chromosomes as the fertilized egg. In 1883 Edward van Beneden suggested that in the two-chromosome variety of Ascaris, one chromosome came from the sperm and one came from the egg. When the sperm cell entered the egg, a chromosome formed in its nucleus. A single chromosome also formed in the nucleus of the egg. The nuclei containing each chromosome moved to the center of the cell and came to lie opposite each other in close opposition. Then, each chromosome appeared to split just prior to division. The same thing was observed in the four-chromosome Ascaris, where each parent contributed two chromosomes.

A few years later, about 1888, two workers, Theodore Boveri and Oscar Hertwig, reported on a special kind of cell division

which produced the gametes. This kind of cell division resulted
in cells with half the number of chromosomes that were found in
the body cells.

The observation of the reduction of the chromosome number
tied in with another observation which had been made as early as
1850. In that year, a worker named Warneck had seen that the
eggs of certain snails had little round bodies attached to them.
At the time no one could offer an adequate explanation of their
origin or function. It was suggested that the little bodies called
by many "corpuscles polaire" (polar bodies) were cells. This was
a little hard to take at first by those who were trying to convince
doubters that cells reproduced by equal division. A clue to the
nature of these polar bodies was offered when it was observed
that fertilization stimulated the production of these tiny bodies in
some organisms. Ascaris came through again. The whole sequence
of events was pieced together with the help of their chromosomes.

It was shown that the polar bodies were cells and contained the
excess chromosomes which were cast off in the special reduction
division which produced the gametes. The same kind of reduction
division was seen to occur in the production of sperm cells with
one important difference—the actual cell division in the sperm cell
production was an equal division. The cell divisions in the egg
cell were unequal divisions resulting in a relatively large egg
cell which retained most of the protoplasmic substance. The behavior of the chromosomes in the production of male and female
gametes was the same. Surveys of cell division in animals and
plants revealed that the chromosome-halving cell division was
found in all sexually reproducing organisms.

What was the significance of reduction division? One thing
was immediately obvious, gametes had to have a half number of
chromosomes. If they had a full number of chromosomes, the
chromosome number would be doubled with each new generation.
This obviously had not occurred. There was more to it than just
reduction of the chromosome number. In mitosis, the pairs of
chromosomes lined up along the midline of the cell. The chromosomes then appeared to split lengthwise and go to opposite

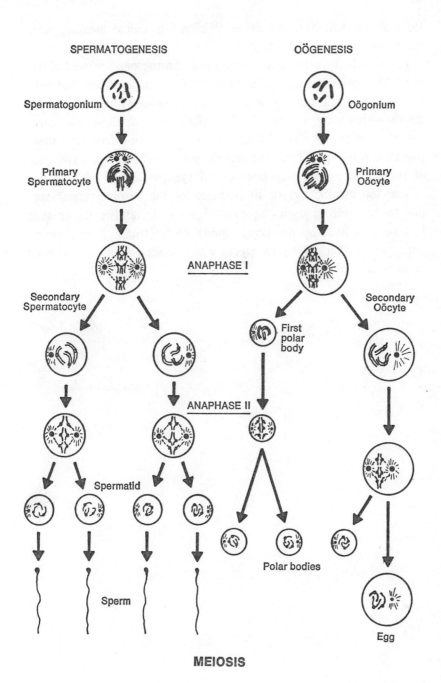

SPERMATOGENESIS

OÖGENESIS

Spermatogonium

Oögonium

Primary
Spermatocyte

Primary
Oöcyte

ANAPHASE I

Secondary
Spermatocyte

Secondary
Oöcyte

First
polar
body

ANAPHASE II

Spermatid

Polar bodies

Sperm

Egg

MEIOSIS

sides of the cell. This reduction division was called *meiosis,* after the Greek word for diminish.

In the reduction division, the pairs of chromosomes moved about until like pairs came to lie very close to each other, actually touching. Further investigation led many to believe that the chromosomes which came together were exactly alike in shape and size. Some investigators offered that one pair had come from the male parent and one pair from the female parent. This coming together of the chromosomes was later called *synapsis.*

The full significance of this "dance of the chromosomes" was not to be realized until the dust was blown off the paper that had been written by the gentle monk from Brunn. It is perhaps appropriate that Mendel's work was "rediscovered" as a new century started.

5. THE GARDEN REVISITED

When Darwin died in 1882, he was buried in Westminster Abbey near kings and scientific greats, such as Sir Isaac Newton. To be buried in Westminster Abbey is an honor accorded to very few people other than royalty and military heroes. That Darwin was buried in the Abbey is testimony to the greatness he had achieved.

The honors received in life and death could not change the fact that his life's work, natural selection, was in serious trouble. The trouble was not in the belief of the mutability of species. By the time Darwin died, most biologists believed that species evolved somehow. However, none of Darwin's followers had been able to offer a satisfactory explanation of how variations happen and how the variations are inherited. Pangenesis was a desperate, if well-thought-out, attempt to save natural selection. There was no experimental evidence to uphold pangenesis. Despite this, there were many who held on to it simply because there was nothing else to believe in.

Following the publication of *Origin of Species,* there gradually emerged two major ideas of the evolution of species. One group believed in Darwin's idea of gradual, continuous change. How did this gradual change take place? They didn't know. The second group came to believe that evolution came about through a series of discontinuous, more or less abrupt variations. How did these abrupt changes happen and how were they passed on to future generations? They didn't know.

The leader of the continuous change group was Darwin's nephew, Francis Galton. They studied the prevalence of certain characteristics in populations of animals and plants. From these studies, they computed volumes of statistics which were supposed to prove continuous variation. Because of their use of mathematics and statistics, they came to be known as the "biometrical school." Most of the members of this group thought, as Darwin did, that the environment could cause changes in organisms which could be passed on to future generations. Many of the biometricians limply held on to pangenesis, in spite of Galton's rabbit experiments.

Galton believed that mathematics was the tool which would eventually find the answers to questions of variation and heredity. He was convinced that everything could be measured mathematically. He even devised a system to measure beauty. He proposed an idea called "The Law of Ancestral Inheritance." According to this law, every individual inherited exactly one half of his characteristics from his parents, one fourth of the remaining characteristics from each grandparent, one eighth of what was left from each great-grandparent, and so on.

Discontinuous variation had many followers. The best known was William Bateson. Bateson became interested in problems of evolution while he was in college. In graduate school, he became convinced that much more had to be found out about heredity if evolution was ever to be understood. Furthermore, Bateson believed that specific, measurable information supported by evidence had to be found.

Bateson was more than a talker. He worked very hard. He went to every country fair where there were exhibits of livestock and cultivated plants. He talked with farmers, animal and plant breeders, gardeners, and anyone else who might be able to tell him something about characteristics and how they are inherited. He went through collections of preserved animals in museums. His travels took him as far as Turkey. In the course of all this, he became convinced of the idea of discontinuous variation. All of this was still not enough for him. He felt that he had no evidence to support his ideas.

He collected his ideas in a book, *Material for the Study of Variation.* Bateson did not make any money from the book, but writing it did help him to get elected to the Royal Society. In speeches to various groups such as the Royal Horticultural Society, he pleaded that general comments about evolution and heredity were not enough. Specific information was needed. Bateson continued to practice what he preached and carried out plant-breeding experiments. However, the specific evidence he sought eluded him.

In the period following Darwin's death many other workers were engaged in breeding and hybridization experiments. Appeals from men such as Bateson were not the only reason for the increase in hybridization work. The population of Europe was rapidly increasing. Governments were concerned about developing more efficient methods of food production. Research to develop hybrids of plants which would yield bigger crops was encouraged.

Everybody working the field of evolution, variation, and heredity had been influenced by one of the outstanding biological thinkers of the nineteenth century. This man was Auguste Weismann.

In the 1870's Weismann investigated the reproductive cells of coelenterates. Coelenterates are a group of animals which include such forms as jellyfish and hydra. In the course of investigating these animals, he was impressed with the separateness of the reproductive cells in relation to the rest of the body.

He had to give up microscopical work because of serious problems with his eyesight, but remained active and wrote several essays on the significance of the work being done in cytology, embryology, and inheritance.

Where Darwin was concerned with how organisms changed to

produce new species, Weismann was more concerned with the stability of inheritance. Weismann could not go along with pangenesis. However, he wondered a great deal about what it was that caused offspring to look like their parents. Organisms obviously did inherit characteristics from their parents, and continued to do so generation after generation. If pangenesis was not the answer, then something else had to be. After considerable thought, Weismann offered his own ideas on inheritance. His idea, which he called "The Theory of the Continuity of the Germ Plasm," is probably second only to Darwin's natural selection in the profound influence it has had on biologists. Weismann stated his ideas in an essay which was published in 1892.

In his theory he stated that the germ plasm (reproductive cells) is produced by other germ plasm and is not influenced by the body cells of the organisms. Whether he knew it or not he was somewhat of a prophet when he wrote, ". . . heredity is brought about by the transmission from one generation to another of a substance with a definite chemical and above all, molecular constitution."

Weismann's theory gave a kind of immortality to the germ plasm. The body may die, but the germ plasm was passed on from generation to generation. Whatever was in the germ plasm which gave it its own characteristics was in the germ plasm itself, and not in the cells of the body of the organisms. The substance within the germ cells somehow went out to the body cells and determined what the organism was going to be. This was exactly the opposite of Darwin's pangenesis. According to pangenesis, the body cells determined the composition of the germ cells by sending forth gemmules. According to Weismann, the substance in the germ cells determined the composition of the body cells of the organism. Weismann was fairly certain that the hereditary factors were in the cell nucleus.

Weismann elaborated on his theory in the years that followed. He thought that the "hereditary substance" was somehow associated with the chromosomes. To this hereditary substance, he gave the name *Id*. The only evidence he had to support this was the equal distribution of chromosomes in cell division. Weismann believed that each chromosome had within it all of the substance needed to determine all the characteristics of the individual. He thought that there was sort of a "competition" between the chromosomes. The id was somehow rearranged during cell division. The rearrangements caused the cells to become the different kinds of cells which made up an organism.

Weismann's essays triggered off a lot of controversy, which was inevitable. More importantly, it triggered off more thought and research. Attempts to gather evidence to uphold, or knock down, or modify Weismann's ideas resulted in work which led in almost a straight line to the library shelves holding the works of Mendel.

Experiments done by Theodore Boveri with sea urchins supported some of Weismann's ideas and discredited other parts of Weismann's theory. Boveri found that if he exposed sea urchin eggs to an excessive number of sperm, some of the eggs would be fertilized by two sperm cells. Such doubly fertilized eggs would divide in an abnormal way. Four cells formed at the first division following fertilization. Some of the cells had the normal number of chromosomes and others had an abnormal number of chromosomes. Those cells with the normal chromosome number developed into normal, but smaller than usual, sea urchins. Those cells with an abnormal number of chromosomes developed into "monster" embryos.

If each chromosome had all the id needed to make a sea urchin egg into a sea urchin, then it should not matter how many chromosomes were in each cell. That part of Weismann's theory was effectively weakened by Boveri's experiment. On the other hand, the experiment did tend to indicate that the id, determinants, factors, etc., by whatever name, were in the chromosomes.

A Dutch botanist, Hugo DeVries, was very much impressed by what Weismann had to say. He didn't agree with Weismann entirely, but he thought that much of what Weismann had said supported his own ideas on discontinuous variation.

DeVries summarized his views in a book, *Intracellular Pangenesis* published in 1889. DeVries had a way of picking and choosing from other people's work whatever would support his own ideas. From Darwin, he accepted the idea of individual hereditary units. He did not accept the idea of gemmules circulating through the blood. From Weismann, he accepted the continuity of the germ plasm. He did not go along with Weismann's idea of the id all being in one chromosome and repeated in each. DeVries said that determining units, which he called "pangens," could be recombined in different ways in offspring and redistributed to their offspring. DeVries embarked upon a series of plant-breeding experiments to gather evidence for his ideas.

About three years after he published *Intracellular Pangenesis*, DeVries started another series of plant breeding experiments. In 1894 he crossed two varieties of silene plant. One variety had smooth leaves and the other had "hairy" leaves. In the second generation, he obtained 536 plants. Three hundred ninety-two were hairy and 144 were smooth. The ratio was 3 hairy to 1 smooth. In other experiments, he consistently obtained close approximations of 3 to 1 ratios. He observed that some characteristics seemed to be dominant over others.

He reported on some of his results at various meetings, but he did not publish his findings. He was not quite sure, but he believed that the 3:1 ratios he obtained in his experiments proved that characteristics were segregated and randomly distributed.

DeVries carried out his experiments for seven years, and by 1900 he was apparently ready to publish his results. As far as is known, he was finishing his report, when he received something in the mail. It was from his friend, Professor Beijerinck, of Delft, in The Netherlands (Leeuwenhoek's hometown). Beijerinck's letter included the following: ". . . I know that you are studying hybrids. So perhaps the enclosed reprint of the year 1865 by a certain Mendel, which I happen to possess is still of some interest to you." Beijerinck made what was perhaps the understatement of the century.

It is not known exactly how DeVries reacted when he read Mendel's paper, but there is no doubt that he must have reacted strongly. Some science historians think that DeVries may have known of the existence of Mendel's paper before he received the letter from Beijerinck. But if he overlooked Mendel when he wrote his paper, it is understandable. DeVries was only human.

He had every hope that part of his findings would be labeled "DeVries' Law of Segregation." If Mendel's paper had been found one or two years later, this would probably have been the case. Here was a situation similar to Darwin hearing of Wallace's work. DeVries had labored seven years and now the originality had been taken away from him.

At least two other workers were carrying out breeding experiments similar to those of DeVries. These men were Carl Correns in Germany and Erich von Tschermak in Austria. They were not as thunderstruck as DeVries when they came across Mendel's paper. They worked independently of each other but arrived at somewhat the same conclusions. They found Mendel's paper in the course of the usual researching of the literature. In 1881, a German botanist, Focke, had published a bibliography of all work on plant hybridization. In this work, under the heading of Pisum (peas) he wrote that Mendel "believed that he found constant numerical relationships between the types."

Carl Correns was a student of Naegli. There is no evidence, however, that Correns ever heard of Mendel from Naegli. Among other things Correns showed that the endosperm, the fleshy part of certain seeds, was derived from a "double fertilization." This double fetilization is the fertilization of two female nuclei in the embryo sac by one pollen nucleus. It had been observed in lilies some years earlier. From this study, it was determined that the color of the endosperm (most of the seed in peas and beans) was influenced by the pollen, and therefore, an inherited thing. Correns followed up this work on peas with some work on corn.

In going through the literature on plant hybridization, he came across Focke's reference to Mendel. Correns wrote a paper on his work with peas in May 1900. Correns, in his paper, gave full recognition to Mendel. He felt that the Mendelian type of inheritance might apply to many more organisms than peas.

Erich von Tschermak was a grandson of one of Mendel's professors at the University of Vienna. He was mostly interested in breeding more productive food plants. Tschermak also worked with peas and obtained the 3:1 ratio in the second generation. He also carried out a back cross and obtained the one-to-one ratio that Mendel had obtained. Later he came across the Mendel reference in the Focke bibliography.

Both Correns and Tschermak were in no hurry to publish. They were prodded to publish quickly when DeVries sent them his paper on his work with peas. DeVries' paper was much more of a shock to them than Mendel's paper. It is odd that the paper DeVries sent to Correns made no mention of Mendel, although some of Mendel's terms were used. Papers that DeVries sent to other people did have references to Mendel.

DeVries, Correns, and Tschermak are the men usually credited with the "rediscovery" of Mendel. But it was a fourth man who spread the word of Mendel. This man was Bateson. In his enthusiasm for Mendel, he acted somewhat as a posthumous publicity agent.

On May 8, 1900, Bateson got on a train to travel to another meeting of the Royal Horticultural Society. He was going to deliver another speech on how little was known about inheritance. Many people bring something to read when they go on a trip and Bateson was no exception. In addition to his prepared talk, he had brought along some journals which had arrived in the morning mail. In one of the journals, he found a summary of Mendel's laws, and a report of DeVries' acknowledgment of these laws. Bateson later said that this was one of the most dramatic moments of his life. For years he had been crying out for the necessity of specific work in the nature of inheritance if evolutionary problems were to be solved. Everything he had been looking for was in the work of this unknown Mendel. And it had been around for thirty-five years! Bateson incorporated the account of Mendel's work into his speech. In his speech to the Horticultural Society he said that Mendel's work ". . . will certainly play a conspicuous part in all future discussions of evolutionary problems."

He later found, with no little difficulty, a copy of Mendel's paper and was even more impressed with what he read. This was, certainly, what Bateson had been looking for. Individual traits were studied rather than populations, the characteristics were discontinuous, accurate counts had been made of the progeny, and mathematical interpretations had been made.

It is to be expected that not everyone rallied to Mendel and Bateson. The battle between the pro- and anti-Mendelians was long and bitter. Galton was particularly opposed to Mendelism. But the biometricians were now not the only ones who had precision on their side.

Bateson promoted Mendelism with the zeal of a missionary, but the biometricians used the memory of Darwin to uphold their cause. Bateson had a problem in that he did not have a job. Loudly advocating Mendelism and his almost violent opposition to Darwinian gradual evolution was not likely to help him get one. Most of the people who had the power of hiring were older and more likely to hold to older ideas. Nevertheless, he continued to promote Mendelism. Bateson skillfully planned his campaign to show that only Mendelism could save Darwinism. He wrote a book on the subject, *A Defence of Mendel's Principles of Heredity*.

In his book, Bateson said that Mendelism was the greatest thing for Darwin's theory of evolution since Darwin himself. Bateson was not content to just write books. He engaged in a great deal of research intended to point out the universality of Mendel's principles. Before many people knew what was happening, a new science had been started. It needed a name. In a rather lengthy statement, Bateson proposed a name—genetics. In the mountains of work and controversy which was to follow, Mendelian genetics was to be the winner over Galton's biometrics, but Bateson, ironically, was to be a loser.

Soon after the rediscovery of Mendel, DeVries published a book on his ideas of evolution. This book he called *Die Mutationstherie* (The Theory of Mutation). In his book, he put forth the idea that evolution took place through definite, separate mutations.

Much of what he had in his book was supported by observations he had made on a plant called *Oenothera lamarckiana,* the evening primrose. *Oenothera* is an American species which had been brought to Europe.

DeVries came across a patch of *Oenothera* in a field. He saw that among the plants there were two which were noticeably different from the others. DeVries was quite excited by what he saw. Here were actual mutations! He collected samples of the regular and apparent mutant types and grew them in his garden.

He found that many of the mutant types bred true generation after generation. The regular types frequently produced variant types. He felt sure that these results proved his ideas on mutation. He maintained that his results demonstrated that selection had little or nothing to do with the production of new species. The cytologists would have some fun with DeVries' work a few years later.

When cytologists read Mendel's work, many were able to see a significant relationship between Mendel's work and their own. Mendel had proposed that the "factors" segregate and are redistributed in a random manner to future generations.

Many cytologists claimed that if the word chromosome were substituted for factor, Mendel's paper would be a fairly accurate description of what they had seen in the meiotic cell division which produced gametes. This observation tended to strengthen earlier suspicions that the chromosomes were the home of the hereditary factors. In 1902 an American, W. S. Sutton, made some statements on these observations, and in so doing, forever united cytology and genetics.

William Sutton's life was somewhat the reverse of many who came before him. He started out as a research biologist and later went to medical school and became a surgeon. In 1902 he was a graduate student at Columbia University in New York. One of his teachers was C. E. McClung, who was doing some work with a peculiar little chromosome he had observed in a bug called *Pyrrhocoris*. This chromosome was so peculiar it had been dubbed "X." McClung thought it might have something to do with determining whether a *Pyrrhocoris* was a male or female bug.

Sutton used a large grasshopper for his work. This grasshopper, called the "lubber" grasshopper, was suitable because its cells had only eleven pairs of easily seen chromosomes. Sutton concentrated his work on the production of sperm cells.

He wrote two papers on his work. The first was a detailed account of the sequence of events in meiosis in the grasshopper.

In the second paper, he related the behavior of the chromosomes to Mendelian genetics.

He reiterated the belief that pairs of homologous chromosomes come together in synapses. The pairs which come together were always the same size and shape. By this time, it was generally accepted that the chromosomes retained their identity through many repeated cell divisions.

He started his paper with five general statements:

1. The chromosome group of the presynaptic germ-cells is made up of two equivalent chromosome series, and strong grounds exist for the conclusion that one of these is paternal and the other maternal.
2. The process of synapsis . . . consists in the unions of pairs of the homologous members [i.e., those that correspond in size] of the two series.
3. The first post-synaptic or maturation mitosis is equational and hence results in no chromosomic differentiation.
4. The second post-synaptic division is a reducing division resulting in the separation of the chromosomes which have conjugated in synapsis, and their relegation to different germ-cells.
5. The chromosomes retain a morphological individuality throughout the various cell divisions.

He went on to say that the way the paired chromosomes gathered at the center of the cell was purely a matter of chance. The chance position determined whether gametes would receive chromosomes containing hereditary units of paternal or maternal origin. This chance position of the chromosomes resulted in a very large number of possible combinations of hereditary factors in the gametes produced by meiosis. Sutton believed that his observations were cytological proof of Mendel's laws of segregation and random distribution. Boveri had arrived at somewhat the same conclusions at about the same time. Of course, all of this was based on the unproved idea that hereditary factors were in the chromosomes.

Sutton's teacher, McClung, working with the X chromosome, provided more evidence that chromosomes contained the hereditary determining factors. Sutton's work was an extension of some observations made in 1891 by a cytologist named Henking. Work-

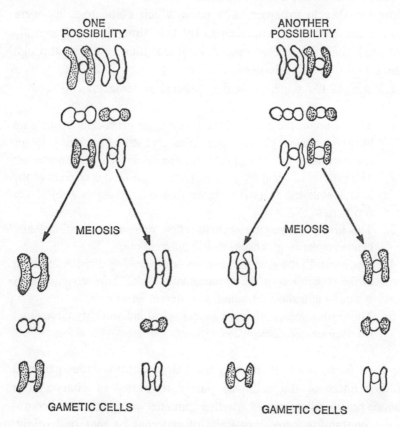

MEIOTIC METAPHASE
SYNAPSIS

ONE
POSSIBILITY

ANOTHER
POSSIBILITY

MEIOSIS   MEIOSIS

GAMETIC CELLS   GAMETIC CELLS

CHROMOSOMES FROM MALE PARENT

CHROMOSOMES FROM FEMALE PARENT

RANDOM DISTRIBUTION OF CHROMOSOMES

ing with meiotically dividing cells of the bug *Pyrrhocoris*, Henking commented on a "peculiar chromatin element." This particular piece of chromatin did not seem to behave as the others did. It did not seem to pair with another chromosome in meiosis. A result of this was that only half the sperm cells had this peculiar X chromosome.

McClung later made observations which suggested that the X chromosome might have had something to do with sex determination. He suggested to Sutton that he study the female *Pyrrhocoris*. This bug was difficult to work with. It had a large number of chromosomes which were difficult to distinguish. Sutton's observations led to the belief that the presence of the X chromosome determined maleness.

It was not until 1910 that it was definitely shown that the X chromosome did have a pairing partner. This chromosome, dubbed "Y," had escaped observation because it is very small in most organisms. E. B. Wilson determined in 1905 that in many organisms the body cells of the male had an X and a Y chromosome and the body cells of the female had two X's. These observations—that the possession, or lack of possession, of a particular chromosome determined so obvious a characteristic as sex, did much to strengthen the chromosome theory.

Bateson had reservations about the association of Mendelism with chromosomes, as did many others. There was good reason to doubt that hereditary factors were in chromosomes. No one had actually identified the factors, much less shown that they were in chromosomes. There was much confusion over just what the hereditary factors were. When men used terms such as id, determinants, pangens, or unit factors, they were referring to particular concepts in their own minds. The same term could mean different things to different men.

William Johannsen, a Danish biologist, did a series of experiments with beans. He was concerned with the inheritance of the size of the seeds. He crossed the largest beans with the largest and the smallest with the smallest. These crosses did not produce beans that were larger or smaller than usual. He wanted to see if this selection would bring about larger or smaller peas than had existed before. The distribution of sizes remained the same.

Johannsen had "selected" beans, but the selection had no effect on the size of the beans. His experiments tended to discredit selection as a mechanism of evolution. Geneticists turned their attention to mutation.

In the process of his bean experiments, he coined the terms

"phenotype" and "genotype." The term genotype was used to refer to the genetic factors present in an organism. Phenotype referred to the appearance of the organism. Johannsen's analysis of his work is not accepted today, but his terms live on. Johannsen also coined the word "gene." This word in time came to replace id, pangene, and the multitude of other terms which had been used to refer to the "something" which determined hereditary characteristics. As it turns out, Johannsen's concept of the gene was somewhat different from our present-day concept.

A biologist, W. Castle, could not go along with the idea that the genes were entirely unaffected by association with other genes. He felt that the genes were "contaminated" to some extent when they came together before segregation. Castle worked with rats, but later suggested the use of the common fruit fly. This little fly was to provide evidence which would eclipse Castle and Johannsen, and firmly establish Mendel as one of the giants in the history of biology.

6. ENTER THE FRUIT FLY

During the 1870's an exotic fruit was introduced into the United States. The banana, imported from Central America, was to be, among other things, the focal point of revolutions in Central America. A little fly which came along with the bananas was to contribute to a revolution in biology.

The tiny fly commonly seen flying about ripe bananas belongs to the genus *Drosophila*. The word Drosophila means "dew lover," but the name is not appropriate. An earlier name, *Oinopota* (wine drinker) is perhaps more descriptive, since the flies seem to prefer fruit that is a little fermented.

Castle knew of the organism called *Drosophila melanogaster* through a friend, C. W. Woodworth. Woodworth was a student of entomology (the study of insects) and was not particularly interested in genetics. Castle, of course, was, and he saw that Drosophila had many advantages for genetic studies.

Drosophila is small and, therefore, takes up very little space in a laboratory. They reproduce rapidly and in large numbers. Under ideal conditions a new generation of Drosophila can be produced every ten days, but the usual time is about two to three weeks. They require little care. Rats and most other animals have to be fed, watered, and kept in bulky cages which require frequent cleaning. Drosophila seem to be content with a little mashed banana, and can be kept in small jars. The waste products of Drosophila are hardly noticeable and the jars require no cleaning unless the geneticist wants to use them again.

As Drosophila came to be more widely used in genetics laboratories it was found to be even more valuable. It has only eight chromosomes. These chromosomes were later found to be an answer to a geneticist's dream. Through Castle, Drosophila came to the attention of a remarkable group of men assembled at Columbia University in New York City. These men, with the help of the immigrant fruit fly, brought new importance to the United States as a leader in scientific research. Before the twentieth century, the great centers of science were in Europe. Very little important scientific work came out of the Western Hemisphere.

Primarily because of the men at Columbia, genetics was to become very much an American science. Their work was to overshadow the work of Bateson in England.

The leader of the "Drosophila group" was Thomas Hunt Morgan. Working with Morgan was Edmund Beecher Wilson. Morgan and Wilson were joined by two men who were undergraduate students when Drosophila came to Columbia. These two men were Calvin Bridges and A. H. Sturtevant. They were enrolled in Morgan's general zoology class. Most professors are not eager to teach the general zoology course. However, teaching the general zoology course enabled Morgan to meet Bridges and Sturtevant.

He was very impressed with the young men and asked them to work with him in his Drosophila laboratory. For a professor to take such notice of undergraduates in a large university is quite unusual. Morgan's feeling about the two men were correct. Bridges and Sturtevant made outstanding contributions, and Morgan never again had to teach the general zoology course.

These four men—Morgan, Wilson, Bridges, and Sturtevant—

were the regulars of the Drosophila group. They were joined by others over the years. Among the more outstanding of these was H. J. Muller. They carried out their work in a very small laboratory which came to be called the "fly room." As workers came to the fly room and left after carrying out Drosophila experiments, they spread interest in Drosophila to other universities. Soon laboratories all over the country reeked from the smell of the various food materials the workers prepared for the flies. These foods or culture media ranged from simple mashed banana to concoctions of syrup, corn meal, molasses, agar, and yeast. As the culture media fermented, pungent odors would fill the laboratory and surrounding areas, much to the displeasure of those not engaged in Drosophila work.

The workers found that half-pint milk bottles were quite suitable for their work. To prepare for an experiment, a layer of culture medium was poured into the bottom of the bottles, and allowed to harden. Parent flies were then placed in the bottle. Wormlike larvae hatched from the eggs and fed on the yeast on the culture medium. Plugs of cotton were placed in the mouth of the bottle to keep the flies from escaping. When the larvae matured into flies, they would be anesthetized with ether, poured out, and counted with the aid of a low-power microscope. The workers would establish ratios of characteristics and try to make some sense from the results. The same procedure is followed in school and research laboratories today. Since half pints of milk are now sold in paper cartons rather than in little bottles, baby food jars have replaced milk bottles as the favorite Drosophila nursery.

Starting from about 1909 and extending well into the 1930's, the Drosophila group utilized the flies in investigating a number of genetic problems. They started with one basic assumption. This was that the genes were associated with the chromosomes. A rival group was working with Bateson in England. Bateson and his followers did not believe that much could be found out about Mendelian genetics from the study of chromosomes. The Drosophila group combined Mendelian-type crosses with cell studies. They attempted to relate the results of the crosses to observations of chromosomes. In the smelly little room, these men accomplished one of the outstanding achievements of the human mind.

Even before much work was done with Drosophila, an apparent contradiction was evident. It was obvious that there were many more characteristics than there were chromosomes. Therefore, if the genes were on the chromosomes, one chromosome would have to contain many genes. Since chromosomes were distributed as units to new cells, the "package" of genes contained in a chromosome could not be distributed independently. This put a strain on Mendel's law of independent assortment. The concept of genes being distributed together on a chromosome was called "linkage."

Correns had noted linkage in 1900. Working with a plant called *Matthiola* he found that some of the characteristics were always found together in apparent contradiction to independent assortment. Other workers made similar observations.

Morgan came across linkage in Drosophila quite accidentally. It was a mutation which enabled him to know that linkage occurred. Mutations were to prove to be quite useful. In 1910 while he was examining a culture bottle full of flies, he noticed that one of the flies had white eyes. The normal eye color of Drosophila is red. The white-eyed fly was a male and apparently represented a mutation. This male was mated with a red-eyed female. The flies in the first generation of this cross were all red-eyed. Morgan could assume that red was dominant over white.

The red-eyed flies of the first generation were then mated to obtain a second generation. The second generation included red-eyed and white-eyed flies in a ratio of about 3 red to 1 white. There was, however, one peculiar thing about the second generation. All of the white-eyed flies were males. About half of the males had white eyes and half had red eyes. All of the females were red-eyed. Now Morgan had to figure out these interesting data.

Morgan reasoned that since the eye-color characteristic segregated according to the sex of the fly, then the genes for eye color had to be on the X chromosome. The XX female, and the XY male scheme of sex determination had earlier been determined to apply to Drosophila.

In expressing his ideas about the inheritance of eye color,

Morgan used the concept of the *allele* which had been developed by Bateson. Bateson had used the term "alleleomorph" to describe genes which controlled the same characteristics. For example, genes for brown hair, blond hair, and red hair would be alleleomorphs. Later on, the term (shortened to allele) was used to apply to genes in the same spot or locus on a chromosome. Bateson also started the use of F_1 and F_2 as symbols for the first and second generations, respectively.

The allele for white eyes was recessive. This was apparent from the ratios. Morgan theorized that only males showed the white-eyed characteristic since they were XY. There was no other X chromosome to house a dominant red allele. Morgan surmised that a white-eyed male could not have white-eyed sons. This was because the male contributes a Y chromosome to the chromosomal make-up of a male. The daughters, however, would all carry the allele for white-eyed since they receive an X chromosome from the male parent. These females do not show the white-eyed characteristic. They have a dominant red allele on one of their X chromosomes. When mated to a red-eyed male, there is a fifty per cent chance that half of their sons will obtain the chromosome bearing the white allele.

All of this was bold supposition on Morgan's part. No one had ever seen a gene, and there were many respected biologists at the time who stoutly maintained that such things as genes could not possibly exist. Assuming that genes did exist, Morgan's association of the eye-color alleles with the X chromosome was the first time anyone had assigned a particular gene to a particular chromosome. The term *sex linkage* was used to describe a genetic characteristic associated with a sex chromosome.

The X and Y sex-determination pattern had been effectively demonstrated by looking at the cells of male and female fruit flies. That males had an X and a Y chromosome and females had two X's could be seen through the microscope. Morgan had very solid evidence on his side. Bridges was to provide even stronger evidence for the chromosome theory.

While Morgan and his men were working with Drosophila, Bateson and his group across the Atlantic Ocean were not idle.

The Drosophila group made use of Bateson's data to strengthen their own ideas on chromosomes. Bateson and a co-worker, Punnett, had observed that certain groups of characteristics tended to be inherited together. Bateson called this "coupling." When characteristics tended not to be inherited together, he called this "repulsion." Bateson and Punnett worked primarily with peas.

Bateson and Punnett found that the coupling groups they established did not always stay together. On occasion they found that certain characteristics were not inherited with their groups. This was very confusing. They used the term "partial coupling" to describe groups which were usually, but not always, inherited together.

Morgan had observed the same kind of thing in Drosophila. After placing the allele for white-eyes on the X chromosome, he found other characteristics which were inherited the same way as eye color. Among these were mutants for yellow body color and miniature wings. (The normal body color of Drosophila is black.) These were determined to be on the X chromosome in the same way that eye color had been assigned to the X chromosome.

In 1910 Morgan carried out a cross between white-eyed, miniature winged flies and white-eyed, yellow-bodied flies. Since the factors were linked, he expected that the offspring would be the same as the parents. However, this was not the case. Some of the flies in the first generation showed a recombination of characteristics. That is, some of them were yellow-bodied with miniature wings. Some of the characteristics had become unlinked!

Morgan saw that his results were similar to those obtained by Bateson and Punnett. He, therefore, knew that his results were not just isolated freaks. He drew on the observations of another man to provide material for another hypothesis. About a year earlier, the Danish biologist, F. A. Jannsens, had observed an interesting thing about chromosomes during meiosis. He saw that homologous chromosomes which had come together in synapsis would sometimes overlap or wrap around each other. To this phenomenon, he gave the name *chiasmatype*. Morgan used it to explain the recombinations.

Morgan called the overlapping of the chromosomes *crossing*

over. He said that when two chromosomes cross over, they break at the point of crossing. The broken pieces of the chromosomes then rejoin. But pieces of the same chromosome do not rejoin. Rather, the pieces of the different chromosomes which were next to each other rejoined. This explained how characteristics which had been assigned to the same chromosome were sometimes inherited separately. Morgan had formed a picture of the genes lying along the chromosomes in a straight line.

He was able to state that the "exceptions" to Mendelian inheritance such as linkage were not exceptions at all. The hereditary factors were distributed randomly and independently. The basic Mendelian pattern was sometimes hidden by linkage, but it was still there.

There was a great feeling of excitement in the fly room as the pieces of evidence began to fit together. The beautiful, simple principles that Mendel had proposed were holding up through one "crisis" after another.

Sturtevant reasoned that if genes were at opposite ends of a chromosome, there was a greater chance that they would be

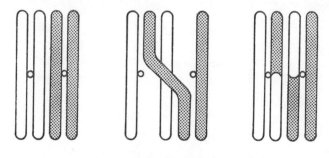

CROSSING OVER

recombined after crossing over. Sturtevant proposed that the percentage of recombinations in a generation of flies was an index of the distance between genes on a chromosome. The greater the percentage of recombinations, the greater the distance between the genes involved in the recombination.

Using the recombination data, Sturtevant actually drew chromosome maps. These maps indicated the relative distance between

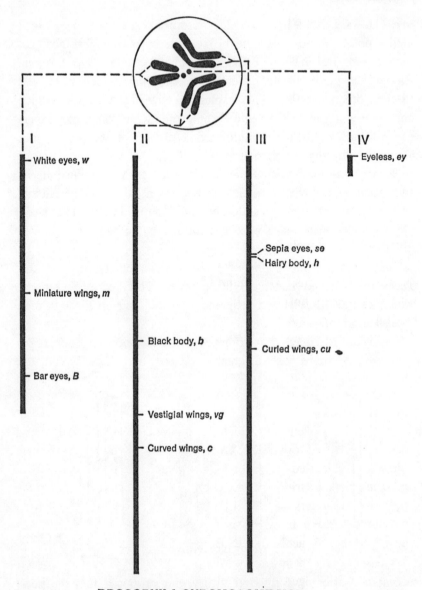

DROSOPHILA CHROMOSOME MAP

genes on the chromosomes. The first chromosome to be mapped was the X chromosome. There was a great deal of data on the X chromosome from sex-linkage studies.

Mapping genes on chromosomes other than the sex chromosomes was more difficult, but over the years it was done. Sturtevant's methods were used by geneticists to map the chromosomes of other organisms.

Sturtevant also used his data to destroy one of Bateson's ideas. Bateson had proposed that a characteristic was recessive because the factor which determined it was just not there. A factor was dominant because it was present. He felt it was possible for a factor to become lost during meiosis and, therefore, not show up.

Sturtevant had observed that recessive mutant alleles, such as these for white eyes, sometimes "mutated back." He had observed that crosses of white-eyed flies sometimes yielded red-eyed progeny. He also observed a new eye color from such crosses, which he called eosin eyes. The only way Bateson's theory could explain this was that somehow an absent gene mutated to a present gene!

Sturtevant proposed that many different alleles could be at a chromosome locus. The mutation was due to a change in the gene itself and not to its being lost. Sturtevant called his idea "multiple alleles." According to this idea, many genes for a characteristic such as eye color existed. In general, the original, non-mutated gene was completely or partially dominant over the mutated genes. Which characteristic showed up was dependent on the interaction of two genes, not one that was present and one that was lost.

Bridges had been very much intrigued with the eye-color business and carried out some crosses of red- and white-eyed flies himself. He carried out a cross between white-eyed females and red-eyed males. Bridges expected to obtain white-eyed males and red-eyed daughters in a 1:1 ratio. This is about what he got, but there were a few flies in the bottle which just should not have been there, according to Bridge's calculations. There were a few red-eyed males and a few white-eyed females. Where did these come from? Bridges thought about it for a while and came up with a brilliant proposal which he later proved with the microscope.

Bridges figured that in the meiotic cell divisions in some flies, the X chromosomes had failed to separate and go to the newly formed cells. As a result, some of the females had received two X chromosomes from the female parent.

Bridges arrived at his hypothesis with the confidence of an

artist daubing paint on a canvas. He proposed that during meiosis in some of the parents an "accident" had occurred. Some of the X chromosomes had failed to separate so that some of the gametes had two X chromosomes. Bridges called this accident *chromosomal nondisjunction.*

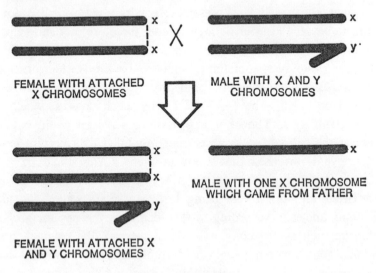

FEMALE WITH ATTACHED
X CHROMOSOMES

MALE WITH X AND Y
CHROMOSOMES

FEMALE WITH ATTACHED X
AND Y CHROMOSOMES

MALE WITH ONE X CHROMOSOME
WHICH CAME FROM FATHER

SEX-LINKAGE

Cell studies with the microscope revealed that flies with XXY were females. The females with white eyes had two X chromosomes with an allele for white eyes on each. Flies with just one X chromosome were males. If the X chromosome happened to have a red allele, the fly had red eyes. Male flies with just an X chromosome and no Y chromosome turned out to be sterile. That is, they could not reproduce.

All of the white-eyed female progeny from the white-eyed female × red-eyed male cross had sex chromosome complements of XXY. Cell studies also verified his thoughts about the chromosomal make-up of the sterile, red-eyed males. They had just an X chromosome in their sex chromosome make-up.

This relating of genetic characteristics to the chromosome complement was overwhelming evidence for the theory that the genes were on the chromosome. The men in the fly room began to call their idea "the chromosome theory."

By 1915 the Drosophila group felt they had enough data to apply all of Mendelian inheritance to chromosomes. Morgan, Sturtevant, Muller, and Bridges wrote a book entitled *The Mechanism of Mendelian Heredity.* The book was a summary of their work with Drosophila. Their basic proposition was that Mendelian genetics could be explained on the basis of genes being on the chromosomes. The book is still regarded as a classic in the field of genetics.

The book did little to convince the doubters of the chromosome and gene theories. In a review of the book, Bateson wrote:

. . . it is inconceivable that particles of chromatin or of any other substance however complex can possess those powers which must be assigned to our factors (genes) . . .

Bateson went on to say that one piece of chromatin was like any other piece of chromatin. He said that since chromosomes were the same chemically, it was impossible for them to contain anything that could determine the different characteristics of an organism.

The work in the fly room did not stop with the publishing of *Mechanism.* Over the next fifteen years the men gathered more ammunition to support the chromosome theory.

In 1921 H. J. Muller delivered a speech in which he more or less summed up what was known about genetics at the time.

He delivered the speech at a symposium on variation conducted by the American Society of Naturalists. Muller also indicated the direction he thought future research should take.

At this same meeting Bateson capitulated and made a speech acknowledging that genes did have something to do with chromosomes. Almost everyone at the meeting knew this was coming but it still caused quite a bit of excitement.

Bateson's admission demonstrated what a fine man Bateson was. He more than anyone else, had made Mendel known to the world. Then he had spent twenty years of his life trying to uphold Mendel. He took the wrong path. The proof of Mendelism be-

longed to others. A lesser man could not have so readily admitted
he was wrong.

Bateson also said that the gene theory had effectively disproved
Darwin's continuous variation as well as DeVries' mutation theory.

There were newspaper reporters at the meeting but they did
not understand much of what was going on. The only thing
that interested them was Bateson's statement about Darwin. Some
newspapers carried such headlines as "DARWIN DOWNED,"
which was a complete misrepresentation of what Bateson said.
The newspaper stories contributed to the passing of laws in some
states which forbade the teaching of evolution in the schools.

As exciting as Bateson's speech was, Muller's remarks were
more important to the development of genetics and evolution than
Bateson's admission or the actions of state legislatures. Muller
stressed the following points in his summary of the then-present
state of genetic knowledge:

1. Genes are definite substances. But it is not known just what sub-
    stance or substances make up the gene.
2. The genes are very, very small. All evidence points to this. Hun-
    dreds of genes had been located on chromosomes which were
    only a few microns long. (A micron is 1/1000 of a millimeter.
    A millimeter is about 1/25 of an inch.)
3. It is not known how genes act. Some investigators have suggested
    that genes control the making of enzymes in the cells. One thing
    that is well accepted is that genes are not the same thing as the
    substances they control.
4. Genes can mutate.

Muller spoke at length on mutation. He said that gene mutation
—that is, variation—is the root of evolution. If men were to find
out how evolution worked, they must find out more about both
inheritance and variation.

Muller wrapped up his talk with suggestions for future research.
He thought that there was much more to be learned from the
study of chromosomes. He emphasized the importance of studying
mutations. He reviewed the past dozen or so years of Drosophila

research in which mutations had been so useful as a research tool. In the years that followed, Muller was to be the leader in mutation studies.

Muller made another suggestion which confused many of those present. He suggested that geneticists could benefit from the study of something which was called the D'Herelle phenomenon. In 1917, the investigator D'Herelle had found a substance which had an interesting effect on bacteria. The substance, called D'Herelle bodies, had been found in association with the bacteria which cause dysentery. D'Herelle found that if this substance were put in contact with the bacteria, the bacterial cells would dissolve. In the process, the amount of bacteria-destroying substance was increased.

Muller felt that the study of D'Herelle bodies could provide a common approach for geneticists, physicists, and chemists. He also thought that these D'Herelle bodies were similar to genes, mainly because they were self-duplicating. The D'Herelle bodies had been referred to as a "filterable substance." This was because they could pass through fine porcelain filters which had been designed to remove bacteria from fluids containing them. The D'Herelle bodies were thought by many to be the same as, or similar to, the "filterable virus" which had been discovered by the Russian, Iwanowski, in 1892.

Muller's suggestion of studying D'Herelle bodies got a mixed reception. Some people thought he was joking. How could these D'Herelle whatever-they-were be of any use to genetics? They couldn't even be seen, not even with the most powerful microscopes. Furthermore, no one who had been working with D'Herelle bodies had even bothered to come to the convention. No one rushed out to start researching D'Herelle bodies in relation to genetics.

Mutation was widely studied, especially by Muller. The Drosophila group and others continued with chromosome studies. Muller had not been content with the small number of natural mutations. He wanted to find some method of inducing mutations. He and other workers had tried exposing Drosophila to high temperatures with limited success. He needed something else and decided to try x-rays.

X-rays had been discovered by Wilhelm Roentgen in 1895, and it was soon evident that they could do more than take pictures of bones. It was found that x-rays had harmful effects on tissues. In view of this finding, Muller felt that x-rays might be useful in inducing mutations in Drosophila.

Muller exposed the flies to x-rays. He found that among other things, there were more lethal genes in the offspring of the exposed flies. The effect of lethal genes had been observed earlier. A lethal gene is one which causes the death of the organism. Lethal genes are recessive. The lethal genes which were linked to nonlethal genes proved to be quite useful in genetic studies.

Muller found that some alleles were more likely to mutate than others. He was able to relate changes in the chromosomes to some mutations. In other instances, no observable changes in the chromosomes could be related to observed changes in the flies.

Other investigators employed Muller's techniques. Muller's work also demonstrated that radiation, which occurs naturally, can cause mutation. When nuclear weapons, such as atom and hydrogen bombs, were developed, Muller's work was anxiously remembered. The radiation resulting from explosions of these weapons can increase the rate of gene mutation among exposed organisms. Concern over this possibility contributed to an agreement between the United States and the U.S.S.R. to stop exploding nuclear weapons in the air.

Many more interesting chromosome investigations were carried out. Bridges worked with a mutant gene which caused the eye of Drosophila to be bar-shaped rather than round. This bar condition occurred in varying degrees of severity. Bridges found that the severity of the bar characteristic depended on the position of the genes on the chromosome. He found that it was possible, as a result of crossing over, for two bar genes to be on the same chromosome. The result was a bar condition so severe that the fly had almost no eye at all. This demonstration, that the position of the genes on the chromosomes made a difference, led to new speculation and confusion as to just what the gene was.

Investigators observed chromosomes which looped, twisted, and turned in on themselves. Most of these chromosome abnormalities were related to observable characteristics of the organisms. Some investigators found extra sets of chromosomes in the cells of some organisms. In addition to Drosophila, the corn plant proved to be useful for genetic and chromosome studies. Peculiar little knobs and protuberances on corn chromosomes were used as "landmarks" to locate genes. Nevertheless, chromosome study was still difficult because of their small size.

The investigators of chromosome abnormalities found DeVries' *Oenothera lamarckiana* to be particularly interesting. The cells of *Oenothera* showed all kinds of peculiar chromosome behavior. Some had extra sets of chromosomes, and some were seen to form loops and circles. The reproduction of *Oenothera* was shown to be very complicated. DeVries' "mutants" turned out not to be mutations at all. The various types of *Oenothera* were due to the inheritance of complex chromosome arrangements. Even though DeVries had been wrong in thinking he had true mutations, his work did point out the importance of mutations in evolution.

By the end of the 1920's, it seemed that Drosophila had provided more information than had ever before been obtained from any organism. Yet, the little fly came through again.

In 1881 Balbiani observed chromosomes in saliva-producing glands of *Chironomus* larvae. Chironomus is a fly similar to Drosophila. These were large chromosomes which had comparatively easily seen light and dark markings.

It was not until 1933 that an investigator found these large chromosomes in the salivary glands of Drosophila larvae. These were actually many duplicated chromosome strands which had come together to give the impression of being single large chromosomes.

These chromosomes were certainly not typical, but their discovery enabled geneticists to clear up a lot of things. The distinctive bands of color were used as landmarks to pinpoint the locations of genes. Chromosome maps were now more than just lines with little marks on them.

If sections of chromosomes were missing, this could be readily seen. Synapsis was dramatically shown to be real when it was seen that sections of whole chromosomes would bulge away from the missing section of a sister chromosome. Various loops, inversions, and other abnormalities of chromosomes could be easily observed.

A PIECE OF A SALIVARY GLAND CHROMOSOME—MUCH ENLARGED

The science of population genetics emerged soon after the rediscovery of Mendel. Population geneticists applied mathematical analyses to studies of changes in populations. Here was the actual application of genetics to problems of evolution.

The arguments between the advocates of continuous and discontinuous variation had really been no argument at all. The picture of evolution which emerged had mutation as its focal point. But it was not the large, "jumping" kind of mutation that DeVries had proposed. The mutations were mutations of individual genes. Each mutation, taken by itself, was small. But the total of the natural selection of many such small mutations over long periods of time resulted in evolution. Cytology, genetics, and evolution were now forever joined.

Both Galton and the Bateson-DeVries group had been right in their own way. Evolution was gradual, and it did come about through mutation.

As the 1930's came to a close, the threat of another world war was becoming a reality. Much had been learned since the dust was removed from Mendel's work. The mechanism of Mendelian heredity had definitely been found in the behavior of chromosomes.

The Drosophila group and others had shown how the genes were distributed. However, they had not been able to show how genes worked and what they were. Disbelievers in the gene theory could still say, "If you can show me a gene and tell me how it works, I might come around to your point of view." As World

War II started, the geneticists still could not show anyone a gene or tell anyone how it worked.

Men who investigated the reactions of matter—the chemists, were to show geneticists the way to the gene. And many a geneticist would wish he had listened more closely to what Muller had had to say in 1921.

7. INTO THE CRUCIBLE

A castle on a river seems an unlikely place for a chemistry laboratory. Unlikely or not, a young chemist named Friedrich Miescher found himself working in a rather dark, dingy room in an ancient castle on the Neckar River in Germany. The year was 1869, and Darwin had just introduced pangenesis. Miescher's choice of work seemed even more unlikely than the place. He had chosen to work on the analysis of pus. His materials were soiled bandages from a nearby hospital.

Miescher was working in the first laboratory which had been set up solely for the study of biochemistry, the chemistry of living things. Men had been interested in what living things were made of for quite some time before Miescher found himself in the castle. Over 2000 years of work had led up to what Miescher and hundreds of others like him were doing.

Chemistry, the study of matter, had been of interest to man long before the rebirth of science during the Enlightenment. Even during the Middle Ages, men called alchemists had investigated matter. The alchemists were not very scientific in the way they went about their work. Many of them spent their entire lives trying to find a way to convert cheap metals into gold. Others looked for potions which would restore youth or prolong life. In the process, they managed to uncover a few bits of information which were useful to the real chemists who followed them.

Alchemists identified and described some of the elements—the building blocks of matter. They devised picturesque symbols to represent what they thought of as elements. Some of the elements known by the alchemists were copper, iron, lead, gold, sulphur, and phosphorus. As might be expected, many of the concepts of matter held by the alchemists were those of Aristotle.

Aristotle thought that everything was made of four basic elements—fire, air, earth, and water. Another Greek, Democritus, proposed that all matter was made of small indivisible particles called atoms. Of course, the ancient Greek concept was nothing like the modern atomic theory. To Aristotle, the various qualities he assigned to matter—dry, cold, wet, and hot—combined in various ways to form the basic elements which in turn made up all matter. As the alchemists found substances, they related them to the qualities set forth by Aristotle.

Chemistry was one of the fastest-growing sciences of the Enlightenment. The gentlemen chemists had a wealth of information, much of it erroneous, from the alchemists. Starting about the middle of the eighteenth century, men such as Lavoisier, Priestley, Dalton, and many others advanced chemistry from the mystical putterings of the alchemists to the status of an experimental science.

The early chemists identified many gaseous elements such as oxygen, hydrogen, and nitrogen. They found that substances could combine to produce new substances. They found that substances could be broken down or analyzed into the substances which made them up.

If any one person could be called the "father of modern chemistry" it would have to be an English Quaker named John Dalton. In the early nineteenth century, Dalton proposed theories about the nature of matter which led to the present day chemical concepts.

Dalton proposed that the smallest chemically acting part of an element was an atom. He said that atoms of a particular element were distinctive because each atom of a particular element weighed the same. There was no technology available in Dalton's time which could be used actually to weigh an atom. Dalton proposed a system of relative weights based on hydrogen, the lightest element. Hydrogen was given a weight value of 1.

Dalton also proposed that elements joined together to form compounds. Furthermore, the elements had to join in definite proportions to form a particular compound. Dalton called the basic unit of a compound a "compound atom." From this idea, the concept of the molecule evolved. The early chemists thought of a molecule as a group of atoms which tended to stay together, but could be broken apart into its elements or changed into another compound with the right chemical treatment. Smaller molecules could combine to form larger molecules.

Early chemists soon realized that living and nonliving materials were made of basically the same matter, but that there were important differences. A "vital" quality was assigned to living material. Chemists were surprised to find that living material was made mostly of only four elements—carbon, hydrogen, oxygen, and nitrogen. They found other elements in living material, but in very small quantities compared to the "big four." They found that living material could not be broken down and analyzed without killing it. They found that, in general, molecules of living material were much larger. That is, they had many more atoms than the molecules of nonliving materials. Chemists applied the name "organic" to those compounds which were produced by living things. Molecules of compounds of nonliving material were "inorganic." This idea had to be modified when chemists found that they could make some of the organic compounds in glass containers in their laboratories.

In 1838 a chemist named Mulder gave the name protein to a prevalent class of organic substance. In 1852 a German chemist, Rudolph Wagner, published a chemical handbook in which he said living material was composed of three basic substances, "carbohydrates, proteins, and lipoids."

So, in 1869, when Miescher was picking through the smelly bandages in his room in the castle, he had a good deal of information to draw on. It was known that atoms of elements combined in definite proportions to make molecules of compounds. No one knew what made the atoms come together or what kept them together. Some sixty-five elements were known, and a Russian named Mendeleyev had just published a book in which he stated that all elements could be arranged in a regular or "periodic" order based on their characteristics. Elements were represented by symbols. The concept of atomic weights was well accepted.

Even though not much was known about atoms and how they came together to form compounds, Miescher had at his disposal many techniques of analysis. The addition of certain chemicals or heat could cause the breakdown of compounds into constituent parts. Chemists of the time could determine the proportions of the elements which made up a compound with a fair degree of accuracy.

Biochemistry had just emerged from organic chemistry when Miescher started his work in the castle, which was part of the University of Tubingen. The nature of proteins, carbohydrates, and lipoids was beginning to be understood. Miescher had been urged to go into biochemistry because his teachers felt that it was the coming thing. At the time, the study of the chemistry of living things was called "histochemistry" because it dealt primarily with the chemistry of tissues. Miescher was aware of the work being done by the cytologists. He had some idea of the cell as being a very important concept in biology.

Miescher chose pus as his object of study because it contains large numbers of white blood cells. Miescher considered the white blood cell to be the most basic kind of animal cell. His project was to analyze the chemical make-up of the white blood cells.

Miescher washed the material off the bandages with salt water. He found that the salt water caused the white blood cells to swell and form a mass. He added a solution of Glauber's salts (sodium sulphate solution) which separated the cells so that they settled out to the bottom of the container he was using.

He extracted the cells with an alkaline substance. He found that a small amount of acid would cause the formation of a gummy substance, which would dissolve in alkalis but not in acids or water. He thought that this substance came from the nucleus of the cells.

His next job was to separate nuclei from the cytoplasm of the cells. He did this by dissolving the cells in dilute hydrochloric acid. This separated out the nuclei, but upon microscopic examination, he found that bits of cytoplasm still clung to the cells. This was not good. Miescher had wanted to be sure that the substance he was trying to identify came entirely from the nuclei.

He treated the cells with an extract of hydrochloric acid taken from a pig's stomach. This treatment resulted in "clean" nuclei free of cytoplasm. But he did not know that it was an enzyme, pepsin, present in the stomach material, that did the job.

Miescher repeated his previous treatment and obtained the same gummy substance as before. He analyzed it and found that it contained nitrogen, sulphur, and phosphorus. It was particularly rich in phosphorus. Since this gelatinous substance came from the nucleus, he called it *nuclein*. He was sure that he had identified a new substance.

Miescher's professor, F. Hoppe-Seyler, delayed the publication of Miescher's paper on nuclein for two years. Hoppe-Seyler had his doubts as to whether nuclein was a new substance at all.

Miescher later extracted the same substance from the sperm of the Rhine River salmon. The only suggestion he made as to the possible significance of the nuclein was that it might throw light on the role of phosphorus in the growth of organisms. Some years after he published this paper, he suggested that nuclein might have something to do with fertilization, but Miescher later abandoned this idea.

Miescher's work caused no particular excitement. He was just another biochemist who had extracted another substance from the material of living things. During Miescher's lifetime most of the attention of scientists and the interested public went to Charles Darwin.

That nuclein was something from the nucleus of the cell did not escape the notice of some of the pioneer cytologists. Hertwig, in 1884, stated briefly that he thought nuclein was important in fertilization and was responsible "for the transmission of hereditary characteristics."

Wilson, in 1895, said that since nuclein was associated with chromatin, which was in the nucleus, ". . . we reach the remarkable conclusion that inheritance may, perhaps, be effected by the physical transmission of a particular chemical compound from parent to offspring."

The views of Hertwig and Wilson were exceptions. Miescher himself doubted the association of nuclein with chromatin. Miescher expressed the views of most people when he said that if any class of substances had to be the hereditary substance, it was protein.

By the end of the nineteenth century, proteins were known to be very large molecules. The atoms in proteins could assume an almost countless number of forms. Even with what little was known about the chemical basis of heredity, it was generally assumed that it was complicated. The complex nature of protein molecules, and the prevalence of protein in living things made proteins the most likely candidate as the chemical stuff of heredity. When more was learned of the chemical composition of nuclein, it was found that it was very simple in comparison to proteins. This finding pushed nuclein out of contention as the possible hereditary substance. This was the case with a generation of geneticists after the rediscovery of Mendel.

The chemical nature of proteins attracted most of the attention of biochemists. They were interested in protein as the main ingredient of living things. During the last thirty years of the nineteenth century, while Flemming, Hertwig, and others were finding out a great deal about how cells divide, biochemists were finding out much about proteins. There was not a great deal of communication between the cell biologists and the biochemists.

The German biochemist, Emil Fisher, contributed much to the knowledge of protein. Among other things, he found the basic molecular subunit of proteins. These "building blocks" of proteins he called *amino acids*. Further investigation disclosed that there

were only about twenty or so amino acids which made up all
of the tens of thousands of different kinds of known proteins.
And more proteins were being identified all the time. This was
a little hard to believe at first.

New knowledge about protein helped to clear up the nature
of a particular kind of substance found in living things. This class
of substance had for quite some time been known to be very
important in metabolism—a term applied to all of the chemical
activities which keep an organism alive. By the end of the nineteenth century, these important substances were called *enzymes*.

Enzymes were not always called enzymes. They had been
called ferments, mainly because some of the first ones to be
identified caused fermentation. As might be expected, many were
discovered before they were called ferments, enzymes, or anything
else.

It was determined that fermentation was a catalyzed chemical
reaction. That is, a catalyst was needed to make the reaction occur
at a reasonable speed. A catalyst is a substance which changes
the rate of chemical reaction without permanently combining with
the reacting substances.

In the 1840's after fermentation was shown to be a catalyzed
reaction, a controversy arose over whether fermentation was a
vital (living) or chemical (nonliving) process. As with many such
arguments, the people involved were both a little right and a little
wrong. Whatever the case, no one knew how ferments worked.

By the time a German chemist, Wilhelm Kuntz, introduced
the term enzyme in 1878, it was generally thought that enzymes
were very specific catalysts produced by living things. They made
possible the many chemical reactions that go on inside living
things and keep them alive. Emil Fisher carried out investigations
in the 1890's which suggested that each enzyme was chemically
constructed to catalyze a specific reaction. Biochemists soon realized
that enzymes were very special kinds of proteins which were ab-
solutely essential to life.

At first, most geneticists were more concerned with what the
chemical nature of the gene might be than they were with
enzymes. They had an idea that the genes somehow directed

the chemical activity of the cell. Geneticists of the Drosophila period felt fairly certain that the genes were chemically different from whatever they caused to be produced.

The suspicion that genes were chemical in their action was strengthened by some developments in medicine in the early 1900's. A physician named Archibald Garrod suggested that certain diseases might be inherited. He cited the case of a disease called alkaptonuria. It was not really a serious disease, but it had a peculiar effect. The urine of people with alkaptonuria would turn various shades of red, brown, or black on exposure to air.

Garrod made a study of the family history of people with alkaptonuria. He found that it tended to occur within members of a family, especially among children of first-cousin marriages. This led Garrod to believe that alkaptonuria was caused by a recessive gene.

Garrod demonstrated that alkaptonuria was due to the failure of the body to utilize components of certain amino acids. He felt that the lack of one specific enzyme brought about this failure.

Garrod tried to popularize his idea of "inborn errors of metabolism." He hoped to generate a joint effort of geneticists, chemists, and doctors, but did not succeed. Bateson referred to him on occasion. Bateson thought that hereditary factors operate by causing the production of enzymes. He did not know much about biochemistry so he could do nothing more than talk about genes and enzymes. Muller reiterated the views of Garrod and Bateson, especially in his 1921 speech to the American Naturalist Society. But for the most part, geneticists continued to count flies and look at chromosomes. Most geneticists of the time had no training in biochemistry.

Despite Muller's insistence that geneticists pay more attention to biochemistry, it was not until the 1940's that any significant work was done which combined the biochemical and genetic approach. One of the many geneticists who saw the need to approach genetics problems through biochemistry was George Beadle.

Beadle was a corn geneticist. He did some work at the California Institute of Technology with Morgan and Sturtevant. While there, he became interested in the pigments which cause eye color in

Drosophila. He went on to Paris to work with Boris Ephrussi who had developed some interesting techniques for studying Drosophila eye color. These techniques involved transplanting the developing eye region from one Drosophila larva to another. Beadle and Ephrussi found if they transplanted a larval vermilion eye to a larva known to be developing into a red-eyed fly, the resulting fly would have red eyes. In other instances, they found the larva would develop eyes the color of the transplant.

On the basis of their work, they stated that much more information was needed on the biochemical action of specific genes. In 1941, Beadle teamed up with a biochemist, Edward L. Tatum, of Stanford University in California.

They needed an organism for their work. Drosophila was not suitable. The rate of mutation, even with radiation, was too slow. At the suggestion of Morgan, they selected the pink bread mold, *Neurospora*. The stated purpose of their experiments was to determine if genes controlled certain biochemical reactions, and if so, how.

Neurospora was very well suited for genetics work. The nuclei produced by meiosis are contained in a saclike structure in the order they are produced. With the aid of a low-power microscope the nuclei (called spores) can be individually teased out for further study. New generations are produced rapidly and in large numbers. Neurospora ordinarily grows on bread. In the laboratory it is usually grown on a gelatinous substance called agar. Food materials can be added to the agar as needed. Neurospora is able to synthesize everything it needs (except one vitamin) from certain inorganic salts and carbon. It was known that enzymes in Neurospora enable the mold to synthesize its needs.

Beadle and Tatum exposed cultures of Neurospora to x-rays. Single spores were then teased out of the sac and placed, one each, on agar containing the "complete medium." The complete medium contained all of the materials needed by Neurospora, those which it ordinarily can synthesize. Even a mutant could grow on this medium. The single spore soon grew into a mass of mold which repoduced many thousands of spores. Since they all came from the same spore, they were all genetically the same.

The workers would then determine if they had a mutant. They would do this by placing a spore in a test tube of "minimal media," that is, a medium which contained only the basic inorganic materials and carbon source from which Neurospora ordinarily synthesizes its needs. If the spore did not grow, they knew they had a mutant.

Then spores were placed, one each, in test tubes containing specially prepared media. Each test tube contained a medium lacking in one material vital to Neurospora. If the mold failed to grow in a particular tube, Beadle and Tatum knew that the mutation involved the inability to synthesize the material lacking in that tube.

Beadle and Tatum thought that mutant forms lacked a gene which controlled the making of a particular enzyme. In their report they claimed they had demonstrated that genes act by directing the synthesis of enzymes. They stated further that one gene controlled one enzyme. This idea soon became known as the one gene-one enzyme hypothesis.

Beadle and Tatum were the first to indicate the biochemical action of genes. There were still many unanswered questions. "How" was, of course, a big question. Another was the biochemical identity of the gene itself. Genes were not enzymes. The enzymes which genes controlled had not been found in the nucleus where the genes reside. By the 1940's there were many who thought that the way to the gene had been pointed out by an English bacteriologist, Fred Griffith.

In the 1920's Griffith was working with the bacteria which cause pneumonia. These bacteria belong to a group called Pneumococcus and Griffith was working with two different strains. One strain had a smooth appearance when it grew on agar, and the other had a rough appearance. The reason for this was that the smooth strain had a coating of slime around it, and the rough did not. The two strains were called S (smooth) and R (rough). Another difference is that the S strain causes pneumonia and the R strain does not.

In the course of an experiment, Griffith injected live R and dead S bacteria into mice. Much to Griffith's surprise, the mice died of pneumonia. Upon examining the blood of the dead mice, he found it full of live S bacteria. Griffith thought he might have been careless and not killed all the S strain. He repeated the experiment, making sure he killed all the S strain. The same thing happened. The mice died from pneumonia and live S strain bacteria were recovered from the blood of the dead mice.

Griffith tried injecting killed S strain alone. The mice did not get pneumonia. Other workers followed up Griffith's experiments. Some found that when dead S strain material was put into a container was live R strain bacteria, the R strain cells became S cells.

Here was a case of genetic transformation. It was quite startling to geneticists, almost Lamarckian in concept. Something in the dead S cells was causing the R cells to change to S cells. What was this something? The transformed S cells reproduced more S cells; the change was definitely genetic.

Sixteen years were to pass before the transforming substance in the S strain was identified. This work was done at the Rockefeller Institute in New York City by three men: Oswald Avery, Colin MacLeod, and Maclyn McCarty.

Their plan of attack was to isolate the chemical compounds of the S strain (called type III). They would test each to see if it was the transforming material.

They worked on the slime capsule first. Nothing in the slime capsule effected the transformation. They then turned to the bacterial cell itself. The techniques of purification of the extracted chemical itself were very difficult. After many trials, they isolated and identified the transforming substance. In their 1944 paper, they stated their conclusion as follows:

> The evidence presented supports the belief that a nucleic acid of the desoxyribose type is the fundamental unit of the transforming principle of pneumococcus type III. (Desoxyribose is now called deoxyribose.)

Biochemistry and genetics now had common direction.

Nucleic acids were not newly discovered substances in 1944.

They were none other than the nuclein isolated by Miescher in 1869. Work done by biochemists in following years had shown nuclein to be acid in nature. It was then dubbed nucleic acid. Miescher's methods of extraction had been crude and what he called nuclein was not pure nucleic acid. Miescher had detected sulphur in nuclein. Nucleic acids do not have sulphur.

Biochemists identified two types of nucleic acid. One was originally called "thymus nucleic acid" and the other "yeast nucleic acid." These names referred to types of nucleic acid which had been extracted from thymus gland cells and yeast cells, respectively.

In 1924, Robert Feulgen discovered a very specific color test for what was still called thymus nucleic acid. This was a dye which turned a magenta color in the presence of the nucleic acid. Feulgen tried the test on preparations of cells. He found that only the nuclei of the cells took up the color. This was in all cells, not just cells of thymus glands.

Feulgen's test indicated that the thymus nucleic acid was found only in the nuclei of cells. One of Feulgen's students, M. Behrens found that yeast nucleic acid was generally located in the cytoplasm of cells. On the basis of incomplete chemical analysis, thymus nucleic acid was called "desoxyribo-nucleic acid" and yeast nucleic acid was called "ribonucleic acid." They were generally called DNA and RNA for short.

Suspicions of a genetic possibility for DNA had never entirely faded away. The exclusive assignment of DNA to the nucleus where the chromosomes dwell, excited the imaginations of a few people. But only a few. Some biologists thought that DNA might be a kind of "glue" that held the proteins of chromosomes together.

There was more evidence which pointed to nucleic acid as something which might be of interest to geneticists. This evidence came from work with viruses. Viruses had been discovered to exist in 1895 by a Russian bacteriologist named Iwanowski who was trying to find the cause of tobacco mosaic disease, a disease of tobacco plants.

Iwanowski passed some juice from infected tobacco plants through a porcelain filter, which had pores fine enough to trap bacteria. When Iwanowski exposed the tobacco leaves to the filtered juice, the plant contracted the disease. Since the causative agent was small enough to pass through the filter, Iwanowski called it a *filterable virus.* He was unable to see the virus with microscopes available at the time.

In the 1930's Dr. Stanley Wendell uncovered some startling information about viruses. He isolated the tobacco mosaic virus (TMV). From about a ton of diseased tobacco leaves, he extracted about half an ounce of a white crystalline powder. He demonstrated that a solution of this powder would cause the tobacco mosaic disease. The white powder was not alive, yet when it was applied to the tobacco leaves, it reproduced itself. It was soon apparent that viruses required living cells in order to carry out the functions of living things. Viruses were thought to be on the borderline between living and dead material.

Following Wendell's work, and idea of the chemical nature of viruses was determined. Viruses had been thought to be proteins. Analysis of the TMV revealed that it was composed of protein and a small quantity of nucleic acid. Other viruses were analyzed and it was shown that all viruses are made of nucleic acid and protein.

The development of the electron microscope in the 1930's and 1940's enabled biologists to see viruses. The electron microscope also helped cytologists to get a better picture of the structure of cells.

A rather interesting kind of virus had been accidentally discovered by Felix D'Herelle in 1917. He had mixed cultures of viruses and bacteria together. He wanted to see what kind of disease the mixture might cause when injected into animals.

When he examined the mixed culture the next morning, he was quite surprised to see that the bacteria were gone. The viruses had apparently destroyed the bacteria.

Further investigation uncovered more viruses which destroyed bacteria. D'Herelle called these viruses *bacteriophages,* which means bacteria-eaters. Many biologists, including H. J. Muller, called them D'Herelle bodies.

Following D'Herelle's discovery, there was a great deal of interest in bacteriophages. For the most part, the men who worked with bacteriophages were interested in the diseases caused by viruses and not in the genetic possibilities of viruses.

In later years when it was learned that viruses are made of protein and nucleic acid, some geneticists became interested in bacteriophages. Many remembered Muller's suggestion that bacteriophages might be similar to genes. Some geneticists thought bacteriophages actually were genes. Most of the workers concentrated their attention on the protein portion of the bacteriophages because they still thought that, if and when the hereditary substance was ever found, it would turn out to be a protein. After the work of Avery, MacLeod, and McCarty, however, nucleic acid started to attract more attention than protein as the possible hereditary substance.

A controversy, soon arose between the advocates of protein and the advocates of nucleic acid as the genetic substance. Some experiments carried out by the Americans, A. D. Hershey and Martha Chase, in 1952 tended to shift opinion to the side of nucleic acids.

Hershey and Chase used techniques of "labeling" with radioactive atoms. They worked with a bacteriophage called T_2. The T_2 virus is shaped somewhat like a tadpole. It has an inner core of DNA and a "coat" of protein. This bacteriophage attacks a bacteria called *Escherichia coli* (*E. coli,* for short). This bacteria is commonly found in the large intestine of animals, including man. The viruses were known to attack the bacteria by entering the cell in some way. The viruses then use the material of the bacteria to make new viruses. The attacked bacterial cell would literally explode, releasing thousands of new virus particles.

Hershey and Chase wanted to know if it was the protein or DNA portion of the virus which entered the cell. The two workers cultured *E. coli* in a medium containing radioactive sulphur and radioactive phosphorus. Sulphur was known to be a component of the protein coat and phosphorus was known to be a component of the DNA. The bacteria absorbed the radioactive material. Then they exposed the radioactive *E. coli* to T_2 bacteriophages.

The next step was to expose nonradioactive bacteria to the now radioactive bacteriophages. In the process of using the bacteria to make more of themselves, the newly formed bacteriophages would be radioactive.

When the newly formed bacteriophages were extracted, it was found that the DNA contained radioactive phosphorus. The protein coat did not contain radioactive sulphur. These results indicated that it was the DNA and *not* the protein which entered the bacterial cell. Later work revealed that some viruses had an inner core of RNA rather than DNA.

Further work, aided by the electron microscope, revealed the actual way in which the viral nucleic acids entered the bacteria. The tadpole-shaped T_2 bacteriophage attached to the wall of the bacterium by means of its "tail." The DNA was actually "injected" into the bacterium. The protein coat remained outside the cell. It was observed that approximately fifteen minutes after the viral DNA was injected into the bacterium, the bacterial cell burst, releasing thousands of new virus particles.

The work with pneumococcus and bacteriophage definitely pointed to DNA as the possible hereditary substance. Some people realized twenty years too late that Muller was not joking when he suggested that geneticists could learn something from working with D'Herelle bodies.

Bacteria and viruses had indeed proved to be quite valuable to geneticists. In 1946 a latter-day genius named Joshua Lederberg teamed up with Edward Tatum to demonstrate some startling things about *E. Coli.* These two gentlemen proved, through a brilliant series of experiments, that *E. coli* bacteria could, under certain conditions, reproduce sexually. This was a bombshell. Everybody "knew" that bacteria always reproduced by asexual cell division. The evidence presented by Lederberg and Tatum was overwhelming. Cells of *E. coli* actually did temporarily fuse and exchange chromosomal material.

The world was to hear more from Lederberg. In 1952 he collaborated with N. Zinder to demonstrate something even more startling in the "sexy bacteria." Lederberg and Zinder demonstrated that viruses could carry genetic information from one bacterial cell to another. Lederberg was awarded a Nobel Prize for his work.

REPRODUCTION OF BACTERIOPHAGES

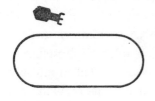

BACTERIOPHAGES BACTERIUM NEAR

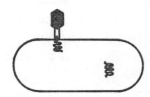

VIRAL DNA ENTERS BACTERIUM
PROTEIN COAT REMAINS OUTSIDE,
VIRUS ATTACHED TO CELL

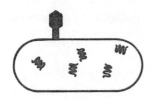

NEW VIRAL DNA MOLECULES MADE

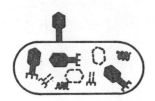

NEW VIRUSES AND VIRUS PROTEIN
COATS ARE MADE

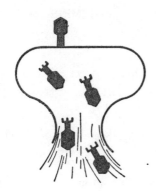

BACTERIUM BURSTS, NEW VIRUSES
ARE RELEASED

It was soon apparent that if DNA was to be positively identified as the hereditary substance, much more had to be known about its chemical structure. In the early 1950's very little was known about how nucleic acids were put together. DNA was known to be composed of a sugar called deoxyribose, a complex of phosphorus and oxygen (phosphate groups), and four molecules called nitrogenous bases. To a chemist, the word sugar means more than what is put into coffee. Sugar refers to a large number of molecules which have a particular structure in common. The bases were called adenine, thymine, guanine, and cytosine. RNA was similar except that the sugar, in its composition, was called ribose and in place of thymine it had the nitrogenous base, uracil.

Deoxyribose differed from ribose only in that it had one less oxygen atom than ribose. The nitrogenous bases were known to be molecules shaped like polygons. The polygons are composed of

ADENINE **CYTOSINE**

GUANINE **THYMINE**

atoms of nitrogen and carbon. Various other atoms are attached to the nitrogens and carbons. There are two general types of nitrogenous bases—the purines and the pyrimidines. The purines are double polygons and the pyrimidines are single polygons. Adenine and guanine are purines, and thymine, cytosine, and uracil are pyrimidines. A group of molecular units consisting of the sugar (ribose or deoxyribose), a phosphate, and a nitrogenous base is called a *nucleotide*. Biochemists had no idea how these small molecules were put together to make the large nucleic acid molecules.

ADENINE NUCLEOTIDE

The evidence that DNA was the stuff of which the gene is made was very strong. If this was the case, the inevitable question loomed—How does DNA act as the genetic material? Before anything could be learned about DNA's possible genetic activity, more had to learned about the structure of this giant molecule. Biochemists began working on the chemical structure of the nucleic acid. Many of these biochemists, such as Linus Pauling, were well known. Others were hardly known at all, except by their immediate friends and family. Among the latter group were a young American named James Watson and an almost-as-young Englishman named Francis Crick. Common interest in DNA had brought the two together at Cambridge University in England in 1951.

Watson had become interested in DNA when he was working for his Ph.D. at Indiana University. His major work had been with bacteriophages. His professor, the microbiologist Salvador Luria, advised Watson that if he expected to find out much about

DNA he would have to beef up an extensive inacquaintance with biochemistry. To that end, Watson obtained a grant of money to study the biochemistry of nucleic acids with Hermal Kalckar in Copenhagen.

Watson was well aware of the genetic possibilities of DNA. He was amazed that in the early 1950's, there were still many geneticists and biochemists who did not seem to realize DNA's significance. Unfortunately Kalckar was more interested in the metabolism of DNA than its possible genetic significance. Failure to see the wider possibilities is unfortunately a frequent problem with scientists who become intensely involved with the details of their own work.

At a meeting of biologists in Naples, Italy, Watson found out more about methods of studying DNA. At this meeting, he met Maurice Wilkins who was studying DNA with the technique of x-ray crystallography. This technique involved passing x-rays through molecules of crystalline substances. X-rays were passed through a sample of the substance and directed onto photographic film. When the film was developed, characteristic patterns were seen on the resulting picture. The patterns were clues to the structure of the molecule. Interpretation of the patterns could give the investigator an idea of how the atoms were arranged in the molecules.

Watson knew less about x-ray crystallography than he knew about biochemistry. He arranged to go to Cambridge to study x-ray crystallography with Max Perutz, one of the leading experts in x-ray crystallography. When Watson arrived at Cambridge, it did not take him long to find Crick. Informal conversation soon revealed common interests. Their decision to work on DNA was complicated by the fact that Maurice Wilkins was working on the same problem. Both Wilkins and Crick had been physicists and had worked with x-rays. Linus Pauling was also known to be working with DNA in the United States. Finding the structure of DNA soon became a race between Pauling, Wilkins, and the team of Watson and Crick. At the time, if anyone had cared to make a bet on who would be first, the wise gambler would have put his money on Pauling. Pauling had established a reputation as the world's leading biochemist.

Soon after Watson came to Cambridge, Pauling published the results of some work he had been doing on the structure of proteins. Pauling, working with sections of proteins called polypeptides, had determined that they were arranged in a helix. A helix is similar to, but not exactly, a spiral. The diameter of a spiral varies at different levels according to a geometric ratio. The diameter of a helix is the same at any point of observation. Pauling had arrived at the alpha helix by constructing a model of the molecule out of objects which resemble children's construction toys. Using what was known about the way atoms came together, and a liberal application of common sense, he had arrived at the alpha helix. Pauling received a Nobel Prize for this work.

Watson and Crick felt that they could arrive at the structure of DNA with the same model-building technique. X-ray crystallography pictures would provide them with much of the information they needed to construct their model. When they learned that Pauling had requested some DNA x-ray pictures from Wilkins, they knew that Pauling was working on DNA. Wilkins thought of an excuse not to send the pictures. But Watson and Crick still thought it best to hurry up with their own work.

X-ray pictures of DNA had indicated that the molecule had a helical structure. The x-ray pictures did not reveal how many strands of atoms the helix had or which atoms were inside or outside the helix.

Some work done by a biochemist named Erwin Chargaff provided Watson and Crick with more information. Chargaff had analzyed DNA from various organisms. DNA had been found in the nuclei of the cells of all organisms which had been analyzed.

Chargaff found that, in all samples of DNA, the amount of adenine always equaled the amount of thymine and the amount of guanine always equaled the amount of cytosine. These data suggested a consistency in structure of the DNA of all organisms. However, if it was DNA which determined the particular characteristics of an organism, then the DNA of different organisms had to differ in some way. Otherwise, if DNA was indeed the genetic stuff, all living things would look exactly the same.

Watson was very much impressed by the work of Hershey and Chase which demonstrated that it was the DNA and not the protein portion of viruses that entered bacteria. This work provided very strong evidence that DNA was the genetic material. This news encouraged the two men to greater efforts. They were probably fully aware that the work of Hershey and Chase would also spur the other DNA workers to greater efforts.

Watson and Crick decided to use Pauling's model-building techniques. They gathered together all the available material for constructing models of molecules. What other hardware they needed, they had made in the University machine shop. Molecular models are basically balls and sticks or other items which represent atoms and the bonding forces which hold the atoms together. Of course, a good deal more was known about how atoms fit together into molecules than was known in Miescher's time. Even though Watson was no biochemist, he had much more biochemical knowledge than Miescher could have ever obtained in his lifetime.

The pieces of metal, sticks, wire, and other items which went into the model could not be put together in just any way. The model had to conform to what was known about angles, distances, and other relationships between the atoms. The job was complicated by certain atoms which could fit into the molecule in any number of ways.

Slowly a picture of the molecule began to emerge from the model-making. X-ray crystallography data indicated that the molecule was helical. A combination of hunch and various data led Watson and Crick to believe that DNA was a double helix rather than a single one. The outer part of "backbone" of the helix was seen to be composed of alternating units of ribose and phosphate groups. The two strands of ribose-phosphate twisted about each other. X-ray data had given them an idea of the diameter of the helix.

Now they had to fit in the purines and pyrimidines. It appeared that the purines and pyrimidines were inside the helix, attached to the sugar molecules. There was room for two nitrogenous

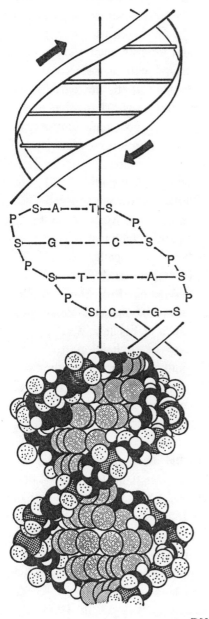

P = PHOSPHATE

S = SUGAR

A = ADENINE

T = THYMINE

G = GUANINE

C = CYTOSINE

 CARBON

 PHOSPHOROUS

 OXYGEN

 HYDROGEN

 BASE PAIRS

DNA

bases between opposite sugars. Two purines were too big to fit inside the helix formed by the two twisted sugar-phosphate strands. Two pyrimidines were too small. A purine and a pyrimidine together fit perfectly. Chargaff's data implied that adenine paired with thymine and guanine paired with cytosine.

There remained the problem of what held the purines and pyrimidines together. This was determined to be a relatively weak bonding force called a hydrogen bond. As the name implies, the bond involves atoms of hydrogen between the purines and pyrimidines.

Since adenine (A) always paired with thymine (T), and guanine (G) always paired with cytosine (C), Watson and Crick proposed that the sequence of nucleotides on one strand of the DNA molecule determined the nucleotide sequence of the other strand. If, for example, one strand had the sequence A–T–C–G–A–A–C–G–C, the corresponding (complementary) sequence on the other strand would have to be T–A–G–C–T–T–G–C–G. They also suggested that it was the sequence of the nucleotide pairs which made the difference between kinds of living things.

It was proposed that in the course of replication, the DNA double helix somehow "unwound" and came apart at the weak hydrogen bonds. Single nucleotides were then exposed along the single DNA strands. Nucleotides which are always floating about in the nucleus would then pair up with the complementary nucleotides on the single unwound DNA strands. When the pairing was complete, there would be two double helices where there had been one before.

After two years of work Watson and Crick had a model of the structure of DNA. They had beat Pauling and Wilkins. Watson and Crick acknowledged that data supplied to them by Wilkins were very important to them. In 1962, when a Nobel Prize was awarded for determining the molecular structure of DNA, it was awarded to Watson, Crick, and Wilkins.

When Watson and Crick published their work in 1953, they opened the paper with the words, "We wish to suggest a structure for the salt of deoxyribose nucleic acid (DNA). This structure has novel features which are of considerable biological interest."

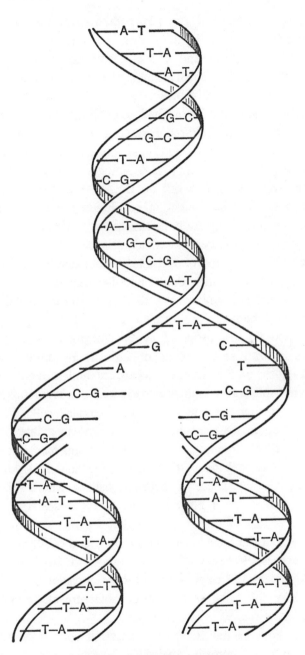

REPLICATION OF DNA

The opening sentences of the paper proved to be one of the grandest understatements in the history of science.

The work of Watson and Crick stimulated an explosion in biology. It started a new science, molecular biology. Never before had so many scientists, all kinds of scientists, leaped to work on one problem. And never before had so much work been done so fast. Attention was now shifted from organisms and cells to the molecules which made up living things. Long-existing lines of division between physics, chemistry, and biology vanished as scientists all over the world went to work on DNA. What they would find would bring on another scientific revolution, perhaps the most significant scientific revolution in the history of man.

Molecular biology encompassed all science in a way that had not existed since the time of Aristotle. Molecular biologists were convinced that if anything could be called the "secret of life" it was the gene and the way it worked. And it looked like they had the gene in their hands in the form of the double helix.

Watson and Crick's proposal of the means of replication of DNA tied in with some details of mitosis. Flemming and the other cytologists of the nineteenth century had thought that the chromosomes split into daughter chromosomes (chromatids). Each chromatid was then seen to move into the new cell. When the cells divided again, two chromatids were seen in a late phase of mitosis. Later work indicated that the chromosomes existed in pairs of chromatids, because the single chromatid had duplicated itself in the "resting phase" between divisions. The term resting phase was seen to be a misnomer, as was the descriptive word "split."

Working from the knowledge that DNA was a major component of chromosomes, Watson and Crick's model of DNA replication offered hope that the duplication of chromosomes could be explained. If so, this was the mechanism of the genetic continuity between generations.

Proof of Watson and Crick's replication idea was provided by studies with radioactively labeled DNA. DNA was labeled with a radioactive nucleotide of thymine. The label was radioactive nitrogen. Experiments indicated that when the labeled DNA duplicated, only one half of the duplicated DNA had the radioactive label. This was shown by exposing photographic film material to preparations of cells containing the labeled DNA.

The DNA molecule is a very long molecule as molecules go. Even the smallest DNA molecules extracted from viruses are composed of many thousands of nucleotides. A knotty question arose. If DNA was the stuff of the gene, then just how much of a DNA molecule was a gene? How many nucleotides make a gene? Would it be possible to identify a single gene in a DNA molecule? A series of experiments done by Seymour Benzer from 1955 to 1961 provided some, but not all, of the answers to these questions. Benzer worked with the DNA of viruses and bacteria. His work bore a superficial resemblance to the Drosophila crossing over and recombination work.

Benzer's experiments involved intensive investigation of a section of bacteriophage DNA. Benzer constructed a "map" based on recombination data. He proposed that single nucleotides were capable of mutation. Benzer also proposed the term "cistron" to describe any particular section of a DNA molecule which carried information for the synthesis of one enzyme. As such, cistron means the same thing as gene. Work on defining the gene in relation to actual nucleotides continues.

The "new biology" may have emphasized the molecule, but there were still a fair number of people who looked at cells and chromosomes. They also contributed in no small way to the revolution in biology.

New techniques were developed for the study of human chromosomes. Human chromosomes are very small and difficult to study. A way was found to spread out the chromosomes so they could be more easily seen. The first information gained from this technique was that the human chromosome number was forty-six and not forty-eight as had been previously thought.

Now that it was easier to count chromosomes, it was found that some human diseases are associated with abnormal chromosome numbers. Among these are "Down's syndrome" which results in retarded mental development. This condition is caused by an extra chromosome. Extra chromosomes result from chromosomal nondisjunction, discovered by Bridges in Drosophila. Some geneticists have even suggested that some criminal behavior may be due to chromosomal abnormalities.

One of the most significant electron microscope observations was the endoplasmic reticulum. This is an extensive network of channels and membranes which extends throughout the cell from the cell membrane to the nucleus. The membranes of the channels were seen to be lined with very tiny spherical particles. These were called microsomes (little bodies).

Some work done by George Palade uncovered some interesting things about microsomes. Palade broke up some cells into a kind of "cell soup," a technique known as homogenizing. The homogenate was then put into a high-speed centrifuge. The centrifuge is a device which has a rotor that revolves at very high speeds. The spinning increases gravitational forces so that tiny cell particles rapidly settle out.

A series of spins separated out nuclei and various other cell components. The fluid left in the test tube after the last spin contained microsomes which were found to be RNA. Some people began to refer to the microsomes as "Palade granules." They are now generally called ribosomes.

Watson and Crick and Wilkins had accomplished a great deal. But the big question still remained unanswered. Just how did the DNA act as the gene—if indeed it was the gene? The proposers of models for DNA action soon made themselves heard. And Crick was one of them.

In proposing his mechanism for DNA-directed protein synthesis, Crick assumed the existence of certain molecules. The only trouble was that these molecules were not known to exist when Crick first did his proposing.

When Watson was working with Crick on DNA structure, he summed up the basic idea of how DNA worked. He wrote the following symbols on a piece of paper: DNA→RNA→protein. Crick elaborated on this theme in a series of papers and lectures around 1958.

Crick's ideas on DNA-directed protein synthesis can be summed up as follows:

1. The DNA in the nucleus directs the synthesis of a strand of RNA. The nucleotide sequence of this RNA strand is complementary to the DNA. Watson referred to this RNA as "template RNA." In later years it was called messenger RNA (mRNA).
2. The template RNA enters the cytoplasm as a ribosome or particle of a ribosome.
3. "Adaptor" molecules in the cytoplasm carry amino acids to the template RNA. Crick went on to propose that there were two types of RNA in the nucleus. One was the ribosomal RNA and the other was the adaptor, which Crick initially referred to as "metabolic RNA."
4. As the adaptor molecules bring amino acids to the ribosomes the amino acids join to form polypeptides and eventually proteins (that is, enzymes).

Later investigations failed to uphold some of the details of Crick's hypothesis. But the amazing thing is that most of his protein synthesis model was upheld by subsequent research.

Studies involving centrifugation of homogenized cells revealed that Crick's adaptor molecule existed. A kind of RNA which was smaller and therefore lighter than other known RNA was isolated. It was known to be lighter than other RNA because it separated out at a slower rate than ribosomal RNA. It was called transfer RNA.

Further cell studies could not hold up Crick's idea that RNA went out to the cytoplasm from the nucleus in the form of ribosomes. Rather, it was thought that messenger RNA left the nucleus and wrapped itself around a ribosome. Even this view was modified in later years when it was thought that the ribosome "rolled" across the messenger RNA. The protein synthesis model generally accepted today is basically the same as that proposed by Crick in 1958.

The soluble or transfer RNA (tRNA) picks up an amino acid and carries it to a ribosome. There was evidence to indicate

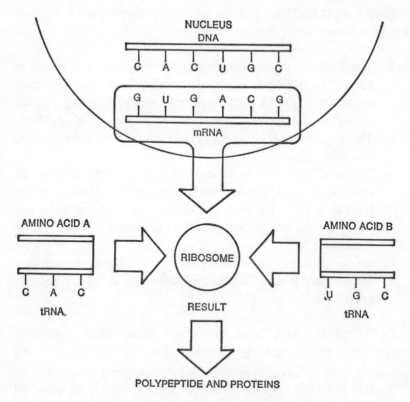

DNA-DIRECTED PROTEIN SYNTHESIS

that there was at least one tRNA for each of the twenty amino acids.

The next question to be answered was a really big one. There are twenty amino acids and only four nucleotides. How could the four "letters" of the nucleotides in DNA determine the sequence of the twenty letters represented by the twenty amino acids which make up proteins? Obviously, one nucleotide for each amino acid was not enough. Combinations of two nucleotides at a time was still not enough. There are only sixteen ways to combine four different things, two at a time. The next obvious possibility was three nucleotides at a time, enough for sixty-four amino acids, well over the twenty known to exist.

Many scientists went to work on this problem which was referred to as "cracking the genetic code." That all kinds of scientists went to work on the genetic code was an indication of how important it was considered to be. If any theme was characteristic of the science of the 1960's, it was unity. There was a unity of scientific effort directed at revealing a unifying principle among all the diverse living things on this planet. George Gamow, an astronomer, had offered a genetic code even before the protein synthesis model was proposed. Gamow's approach to the genetic code did not work, but it was significant that an astronomer had found a genetic problem to his interest.

The race to crack the genetic code was even more furious than the race to find the structure of DNA. It was certainly more publicized than the intellectual battle between Pauling, Wilkins, and Watson–Crick. Everybody knew that solving the genetic code was only a matter of time. It was not a matter of much time at all. By 1962, the code had been effectively cracked.

This work was done by Marshall Nirenberg and J. H. Matthei. They solved the problem with the relatively simple technique of putting the necessary ingredients into a test tube and analyzing what happened. Other workers had become bogged down in complicated and time-consuming recombination studies.

First Nirenberg and Matthei succeeded in putting together a synthetic mRNA consisting entirely of uracil nucleotides. They added the synthetic RNA to a mixture of tRNA, enzymes, all twenty amino acids, and some ribosomes. The enzymes and ribosomes had been obtained by breaking up cells and extracting the materials by centrifugation.

After a period of time, the contents of the test tube were analyzed. It was found that a large polypeptide had been formed. The polypeptide consisted entirely of the amino acid, phenylalanine. Proceeding from the assumption that the genetic code was a triplet code (three nucleotides), it appeared that the RNA message for phenylalanine was UUU.

Other workers, notably Severo Ochoa, who had been a leading contender in the race to crack the code, uncovered more RNA messages. It was found that there were at least two triplet messages for most of the twenty amino acids.

TRIPLET MESSENGERS OF THE GENETIC CODE

AMINO ACID	TRIPLETS		AMINO ACID	TRIPLETS
ALANINE	GCU GCC GCA		ISOLEUCINE	AUU AUC
ARGININE	CGU CGC CGA		LEUCINE	UUA UUG CUU
ASPARAGINE	AAU AAC		LYSINE	AAA AAG
			METHIONINE	AUG
ASPARTIC ACID	GAU ACG		PHENYLALANINE	UUU UUC
CYSTEINE	UGU UGC		PROLINE	CCU CCC CCA
GLUTAMIC ACID	GAA GAG		SERINE	UCU UCG
GLUTAMINE	CAA CAG		THREONINE	ACU ACC ACG
GLYCINE	GUG GGC GGA		TRYPTOPHAN	UGG
HISTIDINE	CAU CAC		TYROSINE	UAU
			VALINE	GUU

Experiments similar to those of Nirenberg, Matthei, and Ochoa, were carried out with nucleic acids from many different organisms. Test after test revealed that the triplet codes were the same in all organisms, bacteria or elephants. The code was universal. Here was the most all-encompassing principle of the unity of all life which had ever been discovered. It was even more universal than the cell theory.

The actual mechanism of mutation was now obvious from the protein synthesis model and the genetic code. A "mistake" involving only one nucleotide in DNA replication, RNA synthesis, or protein synthesis, was enough to constitute a mutation. This mistake is copied in DNA replication and passed on to future generations.

Since the sequence of the nucleotides was seen to determine which amino acids joined to form the protein of enzymes, a change in sequence caused by the loss of only one nucleotide could result in the synthesis of a "wrong" protein. This had been dramatically demonstrated in the case of a disease called sickle-cell anemia. Sickle-cell anemia is a disease of the red blood cells. The red blood cells are unable to carry sufficient oxygen to the body cells. Microscopic examination reveals that the red blood cells of the victims are distorted into a shape that is somewhat like a sickle.

The sickle shape is caused by an abnormality in hemoglobin, the large molecule in blood cells which combines with oxygen and releases the oxygen to the body cells. Hemoglobin is a protein. A molecule of this protein contains over 200,000 amino acids. The hemoglobin of sickle-cell anemia differs from normal hemoglobin in only one amino acid out of the 200,000. It is conceivable that the sickle-cell mutation could involve the loss of one nucleotide which served to change the nucleotide sequence and therefore bring about the change in the amino acid sequence of hemoglobin.

Sickle-cell anemia originated as a mutation in Africa. It has been acted on by natural selection. It would seem that a mutation involving a disease would not be perpetuated, but environmental conditions contributed to the perpetuation of the disease.

In the area of the world where the sickle-cell mutation occurred, malaria is a widespread disease. The organisms which cause malaria enter the red blood cells. Malaria organisms cannot enter sickled cells. Individuals with a mild case of sickle-cell anemia had an advantage in that they could not get malaria. Immunity to malaria gives a greater statistical chance of living long enough to have children than people not immune to the disease. Of course this applies only in malarial areas of the world.

Molecular biology continues to be one of the most active and fastest moving scientific fields. There are still many unanswered questions. Indeed, unanswered questions are what makes a science active.

The gene still has not been exactly defined. Just how many nucleotides make up a gene? Some geneticists have proposed that a gene is the number of nucleotides which direct the synthesis of one polypeptide. Others maintain that a gene is the number of nucleotides which direct the synthesis of one enzyme.

Some work done in 1969 by a Harvard team led by Jonathan Beckwith represents an important step in the definition of the gene. Beckwith succeeded in isolating a section of viral DNA which controls the ability to digest lactose, a sugar. Beckwith selected two strains of bacteriophage which have the lactose gene in common. Beckwith then isolated and "unwound" the incorporated bacterial DNA from the phage particles. Since the lactose gene is the only one the two viruses have in common, the complementary base pairs of this gene were the only ones to combine. Chemical treatment then isolated the base pairs isolating the lactose gene. The lactose gene consists of about 4,000 base pairs.

What tells the DNA molecule when to "unzip" and synthesize RNA? What determines that the "right" portion of the DNA molecule will direct the synthesis of the particular enzymes which are needed by a cell at a given time?

A model by which DNA controls itself has been proposed. This model is called the "operon." According to the operon model, there are two kinds of genes, regulator and structural. Structural genes control the synthesis of protein segments which become part of the structure of the organism. Regulator genes are proposed as genes which can "turn on" or "turn off" the structural genes. This regulatory function may be direct or through polypeptides synthesized by the regulators. At the present time, the operon model is a model which tends to uphold the overall DNA protein synthesis model. Future developments in molecular biology will surely involve the uncovering of evidence which will uphold, refute, or modify the operon model.

All the cells of an organism descend from one cell. All of the cells therefore contain the same DNA. Herein lies one of the most difficult problems in relation to the DNA protein synthesis model. If all of the DNA is the same in the cells of an organism, then why does an organism have different kinds of cells? More information on this problem can be quite significant in the understanding of such problems as birth defects.

There are other practical applications of molecular genetics. Cancer is a disease characterized by cells which cease normal function and increase rapidly to "push aside" normally functioning cells. It would seem that cancer begins when the DNA, which has been directing the normal functions of a cell, somehow changes. The cell is now no longer a nonreproducing, functioning cell, but it becomes a rapidly reproducing cell which no longer carries out the metabolic tasks it did before. Molecular biologists may find out what factors cause the DNA of a cell to change and direct the cell to become cancerous.

Memory has been determined to be a function of RNA synthesis. Learned behavior patterns have actually been transferred from one animal to another. The possibility of "memory" pills or "intelligence" pills is not at all inconceivable.

As more is learned about how the number and sequence of nucleotides relate to specific enzymes, it may some day be possible to actually change the heredity of organisms, including man. The prospect of this kind of "genetic manipulation" has excited the imagination of many people. Molecular biologists do not believe that genetic manipulation will be possible in the near future. Trying to change one nucleotide out of the millions which are in one DNA molecule among millions of other DNA molecules is the ultimate needle in the haystack. However, molecular biology is a fast-moving science. And hardly ten years ago many "experts" declared that space travel would be impossible to achieve in the twentieth century.

EPILOGUE

The mechanism of evolution is now clearly seen in the genetic code. Darwin could never have known that what he was looking for was locked up in a vial of gummy stuff in Miescher's gloomy laboratory. Nor could Miescher have ever realized that what he did would ever in any way relate to the work of the well-known Darwin or to the work of the unknown monk who was vainly waiting for someone to question him about his work. The gentlemen naturalists of the Enlightenment never dreamed that the unity of life so many of them vaguely thought to exist would be pointed out by that curiosity of an instrument called the microscope. The gentleman chemists never thought their work would be of more than passing interest to the men who study animals, plants, and the remains of long-dead creatures.

The work of thousands of men extending over three hundred years contributed to the present scientific revolution. Many lines of scientific inquiry extended from the Enlightenment and proceeded in different directions through a maze of discovery. Some of these lines of inquiry have met in an extremely important unifying principle.

One characteristic of science will always remain. And that is that the solving of one problem always uncovers many more problems. More of the labyrinth still stretches forth. And it extends through all of what man calls the universe. One might ask at this point that when life is found on worlds other than our own, will our genetic code still rate the description of "universal."

INDEX

Darwin, Mrs. Charles, 30
Darwin, Erasmus, 18–19, 20, 21, 24
De Corporis Humani Fabricia (Vesalius), 6
Defense of Mendel's Principles of Heredity, A (Bateson), 95
Democritus, 119
Deoxyribose, meaning of, 134–35
Descartes, René, 8, 9–10
Descent of Man, The (Darwin), 36, 41
Desoxyribo-nucleic acid (DNA), 129–51
  chemical structure of, 134–35
  directed protein synthesis, 146
  location of, 129
  replication of, 141, 142–43, 148
  X-rays of, 136–37, 138
DeVries, Hugo, 91–96, 112, 115, 116
D'Herelle, Felix Hubert, 113, 130
D'Herelle bodies, 113, 130, 132
Discontinuous variation, idea of, 88–89, 91–92
Discourse on Method (Descartes), 9
Dolland, James, 69
Double helix, 138, 140
Down's syndrome, 143–44
Drosophila (genus), 101–17, 125, 126, 143, 144
  eye color of, 104–5
  linkage in, 104
  number of chromosomes, 102
  salivary glands of, 115, 116
DuJardin, Felix, 75
Dumas, 77, 78
Dutrochet, J., 71, 73

Electron microscopes, 130, 144
Endoplasmic reticulum, 144
Enlightenment, 8–18, 21, 118, 119, 152
Enzymes, 124, 126–27, 147
Ephrussi, Boris, 126
Escherichia coli (bacteria), 131–32
Essay on Population (Malthus), 31
Essay on Species (Darwin), 31–32, 35
"Experiments in Plant Hybridization" (Mendel), 43, 60

Fertilization, sperm cells and, 44–45, 66–67, 77–79
Feulgen, Robert, 129
Filterable virus, 130
Fisher, Emil, 123, 124
Fitzroy, Robert, 25–28, 36
Flemming, Walter, 81, 123, 142
Focke (botanist), 93
Fol, Hermann, 79

Franck, Father Frederick, 48
Franklin, Benjamin, 8
French Revolution, 19

Galapagos Islands, 28–30
Galen, 5, 6
Galilei, Galileo, 70
Galton, Francis, 40–41, 116
  opposition to Mendelism, 88, 95
Gamow, George, 147
Garrod, Archibald, 125
Gene-one enzyme hypothesis, 127
Genes
  chromosomes and, 111–17
  concept of alleles, 105, 106, 109, 114
  distributed on chromosome, 104
  Johannsen's concept of, 100
  nucleotides, 149–51
Genetic code, triplet messengers of, 148
Genotype, meaning of, 99
Geology of South America, The (Darwin), 33
Glutamic acid (amino acid), 148
Glutamine (amino acid), 148
Glycine (amino acid), 148
Gray, Asa, 34
Greece (ancient), 2–3, 39, 119
Grew, Nehemiah, 67–68
Griffith, Fred, 127–28
Guanine (nitrogenous base), 134, 140

Haeckel, Ernst, 38
Hall, Chester Moor, 69
Hearn, Samuel, 38
Helix, meaning of, 137
Hemoglobin, 149
Henking (cytologist), 97–98
Henslow, J. P., 24, 26, 27
Hershey, A. D., 131, 138
Hertwig, Oscar, 79, 83–84, 123
Histidine (amino acid), 148
Histoire Naturelle (Buffon), 15
Hooke, Robert, 64, 66, 67–68, 70, 75
Hooker, Sir Joseph, 31, 32, 33, 34, 35
Hoppe-Seyler, F., 122
Human chromosomes, 143–44
Humboldt, Alexander, 24
Hutton, James, 17, 18
Huxley, Thomas H., 35, 36, 38, 39, 64
Hybrids, production of, 46, 49, 50

Id (hereditary substance), 90–91
Intracellular Pangenesis (DeVries), 91–92
Isoleucine (amino acid), 148

Date Due

9-82			

CAT. NO. 23 233   PRINTED IN U.S.A.